£10.95

MATHEMATICAL MODELING
WITH MINITAB®

MATHEMATICAL MODELING
WITH MINITAB®

WILLIAM P. FOX
United States Military Academy
FRANK R. GIORDANO
United States Military Academy
STEPHEN L. MADDOX
United States Military Academy
MAURICE D. WEIR
Naval Postgraduate School

Brooks/Cole Publishing Company
Monterey, California

Brooks/Cole Publishing Company
A Division of Wadsworth, Inc.

Printed in the United States of America
10 9 8 7 6 5 4 3 2 1

Library of Congress Cataloging-in-Publication Data
Mathematical modeling with Minitab.

    1. Mathematical models--Data processing. 2. Minitab
(Computer system) I. Fox, William P., 1949- .
QA401.M3929 1987 001.4'34 87-6368
ISBN 0-534-03546-9

Minitab® is a registered trademark of Minitab, Inc.

Sponsoring Editor: *Jeremy Hayhurst*
Editorial Assistant: *Maxine Westby*
Production Editor: *Phyllis Larimore*
Typist: *Bob Lande*
Cover Design: *Sharon Kinghan*
Cover Illustration: *David Aguero*
Art Coordination: *Sue C. Howard and Michele Judge*
Printing and Binding: *Malloy Lithographing, Inc., Ann Arbor, Michigan*

PREFACE

BACKGROUND AND PURPOSES

Mathematical modeling is the application of mathematics to explain or predict real world behavior. To verify a hypothesized model, often we use real world data. The data are often numerous and "messy," not the "sanitized" variety normally found in text-books. Typically, in such situations it is desirable to have computational support available to analyze the large amounts of data and to eliminate the tedious calculations inherent in the solution or verification of the hypothesized model. The primary purpose of this text is to provide instructions for using inexpensive computational software, for implementation on either micro or mainframe computers, to support the teaching of a wide range of mathematical modeling applications.

THE MINITAB SOFTWARE

Minitab is a versatile inexpensive software package developed by Thomas A. Ryan and Brian L. Joiner of Pennsylvania State University. While its primary purpose is for statistical applications, we have found many other uses as well. It is available for many mainframe computers and for the IBM PC. Further information about Minitab can be obtained from Minitab, Inc., 3081 Enterprise Drive, State College, PA 16801, (814) 238-3280.

We have found Minitab to be the most versatile among inexpensive software packages. While specialized software does exist that is better suited for particular modeling applications, such as the plotting of continuous models, it is generally available only as an individual piece. Minitab is extremely "user friendly"; students enjoy using it and continue to use it in their subsequent course work. It requires no prior programming experience, and students begin to get results almost immediately.

THE SCOPE AND ORGANIZATION OF THIS TEXT

In this text we do not attempt to illustrate all of the many standard applications of Minitab. An excellent reference manual covering Minitab in detail is _Minitab_, Second Edition, by Ryan, Joiner and Ryan (Duxbury Press, Boston, 1985). Our goal is to present only those Minitab commands needed to solve specific modeling applications which we address.

In the text we begin each chapter by describing the new Minitab commands, defining their format statements, and illustrating their use in short examples. We **boldface** the statements when defining a command format, for easy reference. After covering the necessary commands, we conclude each chapter with longer modeling applications that illustrate the use of the commands covered in the chapter. The actual Minitab sessions used to solve the modeling applications will be presented in Exhibits. Exercise sets are provided in each chapter to test the understanding of the Minitab commands presented. Modeling projects are also given in order to exercise the use of the commands in modeling applications.

It is not our intent to teach mathematical modeling in this text. While the text can be used in a "stand alone" mode, it is organized sequentially to provide computer support to the modeling topics covered in <u>A First Course In Mathematical Modeling</u> by Giordano and Weir (Brooks/Cole Publishing Company, 1985). Many of the illustrative examples presented in that text are given here again, and the exercise sets and projects in this manual are designed in support of that text as well.

William P. Fox
Frank R. Giordano
Stephen L. Maddox
Maurice D. Weir

CONTENTS

MATHEMATICAL MODELING
WITH MINITAB®

1

INTRODUCTION TO MINITAB

```
PROMPTS:    MTB>
            SUBC>
            DATA>
            STORE>

NOTATION:   Constants Kj
            Columns Cj
            Matrices Mj

GENERAL SYMBOLS:
            &
            #

ASSISTANCE:
            HELP Commands
```

Background

Minitab is an on-line computing system especially designed for students and researchers with little experience in computing. The system is very easy to learn (the commands are almost like writing English statements), yet it is quite powerful and versatile, with numerous applications covering many fields. A complete description of the Minitab system is given in the Minitab Reference Manual. (For information write to Minitab, Inc., 3081 Enterprise Drive, State College, PA 16801.) A more elementary introduction is given in the MINITAB Student Handbook, Second Edition, Ryan, Joiner and Ryan, Duxbury Press, 1985, which was designed as a supplemental text to be used in an introductory statistics course. We also note that several versions of Minitab exist. In this text we describe Release 82.1. If your version differs, see the Minitab Reference Manual or consult installation personnel to determine the necessary adjustments (which normally are very minor). Appendix A discusses Minitab Release 85.1 which was just made available.

The Structure Of Minitab

Minitab can be thought of as a worksheet consisting of rows and columns of data. The number of columns varies depending on your particular installation, but there are 1000 columns available in Release 82.1. There are, however, some commands which presently can handle no more than 50 columns at a time. Minitab has about 150 commands from which to choose for input of data, performing operations on the data, storing information, and output of result- ing information. The system can store the entire worksheet, a

1

selected portion of it, or any sequence of Minitab commands to be executed at a later time. Release 82.1 also handles up to 1000 stored constants and 100 matrices.

Minitab is capable of operating in either a <u>batch</u> or an <u>interactive</u> mode. In the batch mode, an entire program is submitted for execution. In the interactive mode, execution occurs one line at a time. The interactive mode is generally the most useful for mathematical modeling applications because the modeler typically wants to study the results of one procedure before the next one is determined. In this text we describe only the interactive model; a minimal adjustment is necessary to execute the same programs in the batch mode.

Notation And Conventions

A user begins a Minitab session by following a procedure to access the system (which may differ from installation to installation). Once accessed, Minitab uses one of the following three prompts:

    MTB>
    SUBC>
    DATA>
    STOR>

The MTB> prompt indicates that Minitab is prepared for you to enter a command; the SUBC> prompt indicates that you may enter a subcommand; the DATA> prompt indicates that you may now enter data; and the STOR> prompt indicates that Minitab will store any information you have provided.

Throughout the text we will be describing many Minitab commands and we will adhere to the following notation and conventions:

Kj	where j is an integer, will be used to indicate that a constant (such as 4.2) or a stored constant (such as K4) may be entered.
Cj	where j is an integer, denotes a column (such as C12).
Ej	where j is an integer, will be used to indicate that either a constant (such as K15 or 4.2) or a column (such as C12) may be entered in this position.
Mj	where j is an integer, denotes a matrix (such as M15).

2

We present Minitab commands using the following format:

PRINt the data contained in columns C1 and C2

where the boldface indicates the required portion of the command. The portion of the command not in boldface is strictly optional and is not read by Minitab. For your convenience, we have prepared a Quick Reference Card at the back of this textbook listing the formats for the commands we use, sequenced according to the chapters in which they appear. We now make several important observations.

1. Minitab reads only the first four letters of a command and any arguments. The term <u>argument</u> includes numbers, stored constants, columns, or matrices. Everything else on the line is strictly optional, and for the user's clarification only. For example, the above command can be entered in abbreviated form as:

PRIN C1 C2

2. Minitab is a "free format," in the sense that the order in which the required arguments appear is important but not the spacing. In the example above, C2 does not have to follow C1 by a prescribed number of spaces.

3. With one exception, Minitab does not distinguish between upper-case and lowercase letters--they are treated equivalently. Thus PRIN is the same as **Prin**. The one exception is the NAME command which does honor both upper and lower case characters. We describe the PRINT command later.

Some General Rules For Minitab

There are some general rules that pertain to Minitab that are best presented here:

1. Each Minitab command must start on a separate line.

2. Data must start on the line <u>following</u> the Minitab command used to signal data entry from the READ or SET commands (to be covered). That is, data must not be entered on the same line as the READ or SET command itself. In the interactive mode, a DATA> prompt appears on the line in which data entry may begin.

3. Numbers must be entered without using commas. The number one thousand, two hundred and thirty-four is entered as 1234 not 1,234.

4. Consecutive column numbers (or stored constants or matrices) may be entered in abbreviated form using a dash (minus sign). For instance, Minitab reads C2-C5 as C2 C3 C4 C5.

5. Lines may be continued, if necessary, by using the symbol & at the end of the line to be continued. No character except a blank may follow the & symbol. If your system does not contain the & symbol, then a ++ symbol may be substituted.

6. The symbol # causes Minitab to stop reading the line. Thus, any information appearing after the # symbol is strictly for user clarification. This symbol allows for a convenient method to provide notes and comments to users. The **NOTE** command, to be covered further on, is also available for that purpose. The # symbol can be used on command lines, data lines, or as the first symbol on any line. If it is the first symbol on a line, the entire line becomes a comment.

7. Columns, stored constants, and matrices can be reused any number of times. When they are reused, all previous contents are automatically erased before new results are stored.

The HELP Command

Most of the commands in Minitab have names that are suggestive of their function. Furthermore, the formats for the commands are almost like writing a sentence in English. Nevertheless, while sitting at a terminal you may forget the name of a command that performs a particular function, or perhaps you forget its format. To assist you with this problem, we have provided a reference card for all the commands covered in this text. Minitab also provides an "on-line" HELP facility. To access the facility, simply type HELP after the command prompt MTB> as we now illustrate:

MTB>**HELP**

Minitab then provides instructions to access explanations of the various commands (organized according to the functions performed by the commands), or a particular command or subcommand, or even a general overview of Minitab. For example, if you have forgotten the name of a command, you will be prompted to respond:

MTB>**HELP COMMANDS**

Minitab then provides a list of functions: Input, Output, Plots, and so forth. You type the function of interest and then Minitab explains the different commands that perform that function. Ultimately, you obtain the format of the command you desire. As another example, suppose you remember the name of a command, say PRINT, but forget its format. You may bypass the repetition of the entirety of the above procedure by typing directly:

MTB>**HELP PRINT**

The HELP facility is easy to use and intertactively directs you to whatever information you desire.

2

GRAPHING
CONTINUOUS MODELS

```
DATA ENTRY:        SET
                   GENERATE

OUTPUT:            PRINT

GRAPHING:          PLOT
                   MPLOT
                   HEIGHT
                   WIDTH

DATA TRANSFORMATION:
                   LET

DATA IDENTIFICATION:
                   NAME
                   INFORMATION

STORAGE AND RETRIEVAL:
                   SAVE
                   RETRIEVE
```

In the mathematical modeling process it is often desirable to graph a model in order to gain a qualitative feel for interpreting the model, or perhaps to use the graph for making predictions. In another situation you may wish to test a proposed model against some observed data. It is helpful in such cases to plot the observed data and overlay the data scatterplot on the graph of the model. In this chapter we study the commands necessary to accomplish these tasks with Minitab.

2.1 Generating Values For The Independent Variable:
The SET And GENERATE Commands

Minitab has the ability to plot discrete sets of data. That is, for a finite set of values for the independent variable, Minitab can plot the corresponding values of the dependent variable. The resulting graph consists of a set of Cartesian points rather than a smooth continuous curve. To use Minitab to approximate the graph of continuous models, we generate many values of the independent variable, calculate the values of the dependent variable given by the model, and graph the data points using as large a graph as necessary to display the results. The use of a large graph may be very important in some graphing applications since, like all computers, Minitab rounds the data. Such round-off means the resulting plot may appear choppy unless the values are spread over a relatively large area. We begin our discussion with the two Minitab commands that can be used for generating values of the independent variable, SET and GENERATE. The SET command is preferable and future versions of Minitab may not recognize GENERATE.

SET

In Chapter 3 you will learn how to use the SET command to enter a discrete data set. Here we address the patterned data feature of the command. The format of the command is:

MTB>**SET** the following data into **Cj**

The SET command alerts Minitab that data is forthcoming for storage in the specified column. No data is to be placed on the SET command line. Rather, Minitab responds with a DATA> prompt after the RETURN key; then your data entry may begin. The format for patterned data is:

DATA>**K1:K2(/K3)**
DATA>**END**

These commands tell Minitab to generate values from K1 to K2 in step size K3. In the first line, the part in parentheses is optional. If K3 is not specified (it is optional), Minitab uses a default step size of 1. The END command terminates the data entry and returns the user to the MTB> prompt. Let's consider several examples:

MTB>**SET C3**
DATA>**1:10/.5**
DATA>**END**
MTB>

stores the numbers 1.0,1.5,2.0,...,9.0,9.5,10.0 in column C3.

MTB>**SET C2**
DATA>**1:10**
DATA>**END**
MTB>

stores the integers 1 through 10 in column C2. Note that since no step size is specified, Minitab assumes a default value of 1.

GENERATE

The format for the **GENERATE** command is:

MTB>**GENE**rate values from **Kj** (in steps of **Kj**) to (**Kj**), put into **Cj**

In the above command the portions in parentheses are optional. If only one K is given, Minitab generates the first K integers. If two values of K are given, Minitab generates the integers between the first K value and the second K value. If three K's are specified, Minitab generates values from the first K value in steps the size of the second K value, until the third K value is reached. Let's look at several examples:

MTB>**GENE**rate the first **10** integers put into **C3**

stores the integers 1,2,3,...,10 in column C3.

MTB>**GENE**rate values from **3** to **8**, put into **C2**

stores the integers 3,4,5,6,7,8 in column C2.

MTB>**GENE**rate values from **4** in steps of **0.5** ending at **6**, put into **C1**

stores the values 4.0, 4.5, 5.0, 5.5, 6.0 in column C1.

In summary, a column to store the data must be specified for the GENERATE command. If, in addition to the column, only one constant K1 is given, Minitab generates the first K1 integers. If two constants K1 and K2 are specified, then Minitab generates the integers from K1 to K2 in steps of 1. If three constants K1, K2, and K3 are specified, then Minitab generates values from K1 to K3 in steps of size K2. Here are several examples in abbreviated form:

MTB>**GENE 5 C1**
MTB>**GENE 5 7 C2**
MTB>**GENE 5 2 9 C3**

The first command stores 1,2,3,4,5 in C1; the second stores 5,6,7 in C2; and the third stores 5,7,9 in C3. While the GENERATE command is still available for use on current versions of Minitab, the command is obsolete and is being replaced by the more versatile SET command.

PRINT

Before performing any necessary transformations on the data, it may be desirable to check if Minitab has indeed generated and stored the required values of the independent variable. To see what values have been created, use the PRINT command with the following format:

MTB>**PRINt** the values in **Ej ... Ej**

The PRINT command can contain a mixture of stored constants, columns, and matrices. If it does contain such a mixture, all of the stored constants and matrices are printed out first, in the order they are given, followed by the printing of all the columns. A few rules need to be listed:

1. One or more arguments may be printed.

2. The number of columns printed across the page can be adjusted. See the OUTPUTWIDTH command in the <u>Minitab Reference Manual</u>.

3. If only a single column is printed, it is printed across the page. If more than one column is to be printed at a time, the columns are printed vertically, next to each other. To print a single column vertically, print the desired column together with the empty column.

4. If a number has too many digits to fit in the allowed printing spaces, the number is printed in exponential format.

5. The PRINT command with no arguments is used to restore automatic printing of data. This use is discussed in Chapter 4.

A few examples:

MTB>**PRINt** the values in **C2**

This command prints the values in C2 in a row.

MTB>**PRINt C1 C2 C3 C4**

This command prints the data in columns C1 through C4 as four vertical columns in the specified order, and can be given in abbreviated form as follows:

MTB>**PRIN C1—C4**

Saving And Retrieving A Minitab Worksheet

In many cases we want to save the results of our current Minitab session for future use, or to ensure that all our computer work is not lost in the case of a temporary outage. An entire worksheet can be saved using the SAVE command whose format is:

MTB>**SAVE** (in **'FILENAME'**) a copy of the current worksheet

After the SAVE command is executed, the file contains all the data from the worksheet, including all stored constants, matrices, and column names (we discuss naming columns in the next section). The filename should be chosen appropriately and enclosed in single quotes. For example,

MTB>**SAVE 'ONE'**

saves the current worksheet in a file called "ONE." If no filename is specified, the worksheet is stored in a "default file" whose name can be ascertained from the installation personnel. After using the SAVE command, you may exit the Minitab system. The worksheet ONE may be retrieved later using the RETRIEVE command whose format is:

MTB>**RETRieve** (**'FILENAME'**)

If a filename is specified, Minitab retrieves the file and restores all the data into the worksheet area, including all stored constants, matrices, and column names that existed when the file was originally saved. Again, the filename must be enclosed in single quotes. If no filename is specified, Minitab retrieves the contents of the current default file. Thus, if you should forget to specify a filename with the SAVE command and leave the Minitab system, you can retrieve the default file and resave it, this time <u>with</u> a filename. An example of the RETRIEVE command is:

MTB>**RETRI 'ONE'**

which retrieves the contents of File ONE and places them in the worksheet area.

A Minitab Session Illustrating Data Generation And Filing

A modeler often has the task of analyzing a model someone else has developed. In Exhibit 2.1 we demonstrate the generation of data that will later be used to graph and analyze submodels for the economics of supply and demand. The submodel for demand is given by

$$d = 442413Q^{-1.72}$$

where d is the price for each item when Q units are produced. The submodel for supply is

$$S = 614 - 8.25Q + 0.0321Q^2$$

Notice in the Exhibit that we wish to analyze demand in the interval 125 to 150 in step sizes of 1, whereas supply is to be analyzed from 125 to 225 in increments of 5. The data is printed for purposes of verification, and then saved for future use. (See Giordano and Weir, <u>A First Course In Mathematical Modeling</u>, page 16, for an introduction to the supply and demand submodels.)

2.1 EXERCISES

Save the worksheets created for the following exercises.

1. a. Generate 10 consecutive integers starting at 1, and store them in column C1.

   b. Generate numbers from 3 to 7.5 in increments of 0.5 and store them in column C2.

   c. Print columns C1 and C2.

Generating Minitab Data

```
MTB > SET C1
DATA> 125:150
DATA> END
MTB > SET C2
DATA> 125:225/5
DATA> PRINT C1 C2
 ROW        C1        C2

   1       125       125
   2       126       130
   3       127       135
   4       128       140
   5       129       145
   6       130       150
   7       131       155
   8       132       160
   9       133       165
  10       134       170
  11       135       175
  12       136       180
  13       137       185
  14       138       190
  15       139       195
  16       140       200
  17       141       205
  18       142       210
  19       143       215
  20       144       220
  21       145       225
  22       146
  23       147
  24       148
  25       149
  26       150

MTB > SAVE 'DEMSUP'
MTB >
```

2.2 Transforming And Identifying Data

A mathematical model suggests an existing relationship among the selected variables. In the two dimensional case, the model hypothesizes a functional relationship between the dependent and independent variables. Given values for the independent variable, the function transforms that data to yield the values predicted for the dependent variable. In the last section you saw how to generate the values for the independent variable. To perform transformations you need to understand which algebraic operations and functions may be used in Minitab.

ALGEBRAIC OPERATIONS, FUNCTIONS, AND THE LET COMMAND

Minitab recognizes the normal algebraic operations:

Symbol	Operation
+	Addition
-	Subtraction
*	Multiplication
/	Division
**	Exponentiation

The arguments may be constants, stored constants, or columns. Additionally, Minitab has many available functions. Among the obvious ones are:

> **ABSO**lute value of **Ej**, put into **Ej**
> **SQRT** (square root) of **Ej**, put into **Ej**
> **ROUN**d the values in **Ej**, put into **Ej**
> **LOGE** (log base e) of **Ej**, put into **Ej**
> **LOGT**en (log base ten) of **Ej**, put into **Ej**
> **EXPO**nential of **Ej**, put into **Ej**
> **ANTI**log of **Ej**, put into **Ej**
> **SIN** of **Ej**, put into **Ej**
> **COS** of **Ej**, put into **Ej**
> **TAN** of **Ej**, put into **Ej**
> **ASIN** of **Ej**, put into **Ej**
> **ACOS** of **Ej**, put into **Ej**
> **ATAN** of **Ej**, put into **Ej**

where Ej may be a constant, stored constant or column. Of course, the arguments must be "consistent." That is, for example, you cannot take the absolute value of each element in a column and store the resulting elements as a single stored constant.

Many other functions are available and we discuss some of them in the next chapter when we consider operations that may be performed on columns (such as sums, mean, standard deviation and so forth). To actually execute these functions and algebraic operations, Minitab utilizes the LET command.

LET

The LET command can be used in the following ways:

1. To name and store a constant. For example,

MTB>**LET K2=3.1416**

allows the user to refer to 3.1416 as K2.

2. To access a particular element in a column using a subscript. For example,

MTB>**LET C2(3)=5+K2**

places in the third row of column C2 the number 5 plus the constant stored in K2.

3. To evaluate algebraic expressions. For example,

MTB>**LET C2=C3+C5+K2**

adds the corresponding elements in columns C3 and C5 to the constant stored as K2, and puts the resulting elements in column C2.

The LET command follows the usual order of precedence for calculations. That is, first functions and subscripts are executed from left to right. Next, exponentiation is executed. Then multiplication and division are executed from left to right. Finally, the addition and subtraction operations are accomplished, again from left to right.

More examples illustrating the LET command follow. Assume the value x is stored in column C1 and you wish to calculate:

a. x^2

MTB>**LET C2=C1**2**

b. $2x^2 + 3.5x + 9$

MTB>**LET C3=2*C1**2+3.5*C1+9**

c. $(x^2 + 2)^{0.5}$

MTB>**LET C4=SQRT(C1**2+2)**

In Exhibit 2.2 we demonstrate the use of the LET command to calculate the supply and demand submodels presented in Exhibit 2.1.

Calculating Supply And Demand Values From Formulas

$$S = 614 - 8.25Q + 0.321Q^2 \qquad\qquad d = 442413Q^{-1.72}$$

```
MTB > RETR'DEMSUP'
MTB > LET C3=442413*C1**(-1.72)
MTB > LET C4=614-8.25*C2+0.321*C2**2.0
MTB > PRINT  C1 C3 C2 C4
 ROW      C1          C3         C2         C4

   1      125     109.432      125      4598.4
   2      126     107.942      130      4966.4
   3      127     106.484      135      5350.5
   4      128     105.057      140      5750.6
   5      129     103.660      145      6166.8
   6      130     102.293      150      6599.0
   7      131     100.954      155      7047.2
   8      132      99.642      160      7511.6
   9      133      98.356      165      7991.9
  10      134      97.097      170      8488.4
  11      135      95.863      175      9000.8
  12      136      94.654      180      9529.4
  13      137      93.469      185     10073.9
  14      138      92.307      190     10634.6
  15      139      91.168      195     11211.2
  16      140      90.051      200     11804.0
  17      141      88.955      205     12412.7
  18      142      87.880      210     13037.5
  19      143      86.826      215     13678.4
  20      144      85.791      220     14335.4
  21      145      84.776      225     15008.3
  22      146      83.780
  23      147      82.802
  24      148      81.842
  25      149      80.900
  26      150      79.974

MTB > SAVE'DEMSUP2'
MTB >
```

We begin by retrieving the two sets of values for the independent variable Q generated in Exhibit 2.1 for the two submodels. The corresponding values for supply and demand are then calculated and saved for future use.

Name And Information

Suppose in considering the function $y = f(x)$ we have generated values for x in C1 and the corresponding transformed values for y in C2. In a typical Minitab session many columns may be generated so it is convenient to keep a ledger to remind us where the various columns are stored. The NAME command allows the assignment of convenient names to columns. The INFORMATION command permits inspection of what has been entered on the worksheet.

NAME

The format for the NAME command is:

MTB>**NAME** for **Cj** is **'name1'**, for **Cj** is **'name2'**,...,for **Cj** is **'namej'**

The name assigned to the column must be enclosed in single quotes. The name may contain up to 8 characters. Blanks may not be used at the beginning or end of a name, nor may a single quote be used within a name. Examples of the NAME command are:

MTB>**NAME** for **C1** is **'x'**,for **C2** is **'y'**
MTB>**NAME C3 'WEIGHT' C4 'HEIGHT'**

After naming a column, all output uses the column name instead of the column number. You may refer to a column either by its name or column number. Whenever you refer to the column by its name, the name must be enclosed in single quotes. For example:

MTB>**PRINT 'x' 'y' 'WEIGHT' C4**

prints out the columns C1 through C4 named in the example above.

MTB>**LET C3='x'+'y'**

adds together columns C1 and C2 (term-by-term) and stores the resulting column in C3.

Obviously, a name cannot be assigned to two columns simultaneously. Minitab provides an error message if you attempt to do this. You can change the name of a column by again using the NAME command. Thus the command sequence

MTB>**NAME C1 'x'**
MTB>**NAME C1 'x**2'**

results in C1 being named "x**2." A name you wish to use for a file name should never be used to name a column.

Both upper and lower case letters can be used in a name. When a name is printed in the output, the upper and lower case letters are used. However, if you wish to refer to a column by its name in a command, Minitab considers the upper and lower case letters as being the same, giving you the option to use either.

INFORMATION

If you have forgotten which columns have been used, which names have been used, where the different variables are located, or which constants have been defined, the INFORMATION command is useful. The format for this command is:

MTB>INFOrmation (on **Cj,...,Cj**)

where the column numbers are optional. If none are specified, INFORMATION prints a list of all the columns that have been used, together with their names and lengths (numbers of elements). The command also prints all stored constants and all matrices that have been used. If you do specify columns in the command, then information is given on just those specified columns.

In Exhibit 2.3 we continue on with the supply and demand example. We first retrieve the workfile of Exhibit 2.2 and then assign an appropriate name to each column. Once each column is named, it may be referred to by that name (enclosed in single quotes) or by its column number, whichever is preferred.

2.2 EXERCISES

1. The equation $F = (9/5)C + 32$ is used for converting temperatures given in degrees Celsius to degrees Fahrenheit. Assuming the data stored in column C1 represents Celsius temperatures, use the LET command to create the corresponding Fahrenheit temperatures and store them in a column named FAHR.

2. Using the LET command, store the coefficients of the model

$y = 3x^2 + 5x + 2$ as constants: K1 = 3, K2 = 5, and K3 = 2.

3. Use the LET command and the constants stored in Exercise 2 above to create a column of y values for the model given in that exercise. Assume the independent variable x varies from 0 to 100 in increments of 0.5.

Naming The Columns In The Supply And Demand Example

```
MTB > RETR'DEMSUP2'
MTB > NAME C1'QUANDEM'
MTB > NAME C2'QUANSUP'
MTB > NAME C3'PRICED'
MTB > NAME C4'PRICES'
MTB > PRINT'QUANDEM''PRICED''QUANSUP''PRICES'
```

ROW	QUANDEM	PRICED	QUANSUP	PRICES
1	125	109.432	125	4598.4
2	126	107.942	130	4966.4
3	127	106.484	135	5350.5
4	128	105.057	140	5750.6
5	129	103.660	145	6166.8
6	130	102.293	150	6599.0
7	131	100.954	155	7047.2
8	132	99.642	160	7511.6
9	133	98.356	165	7991.9
10	134	97.097	170	8488.4
11	135	95.863	175	9000.8
12	136	94.654	180	9529.4
13	137	93.469	185	10073.9
14	138	92.307	190	10634.6
15	139	91.168	195	11211.2
16	140	90.051	200	11804.0
17	141	88.955	205	12412.7
18	142	87.880	210	13037.5
19	143	86.826	215	13678.4
20	144	85.791	220	14335.4
21	145	84.776	225	15008.3
22	146	83.780		
23	147	82.802		
24	148	81.842		
25	149	80.900		
26	150	79.974		

```
MTB >
```

2.3   Constructing Graphs

Now that you know how to generate the x values and compute the corresponding y values, you are ready to construct a scatter-plot using Minitab. The basic command to accomplish this task is PLOT, with format:

MTB>**PLOT** y in **Cj** (from **Kj** to **Kj**) versus x in **Cj** (from **Kj** to **Kj**) where specification of the range from Kj to Kj is optional. If the range is not specified, Minitab chooses a scale that spreads the entire data set over the available area. For example,

MTB>**PLOT C2 vs C1**

plots the values in C2 on the <u>vertical</u> axis and the corresponding values in C1 along the <u>horizontal</u> axis. In this case, Minitab will scale automatically since no range is specified.

   Consider next the following example, which plots y = x+2 for values of x between 1 and 10 with step size of 1:

```
MTB > SET C1
DATA> (1:10)
DATA> END
MTB > LET C2=C1+2.0
MTB > PRIN C1 C2
 ROW    C1     C2

   1     1      3
   2     2      4
   3     3      5
   4     4      6
   5     5      7
   6     6      8
   7     7      9
   8     8     10
   9     9     11
  10    10     12

MTB >
```

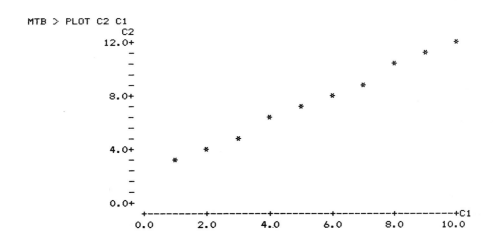

Many times you will want to plot more than one model on a single graph, as in the case of the supply and demand curves. This task is accomplished using the MPLOT command whose format is as follows:

MTB>**MPLOt Cj vs Cj, and Cj vs Cj,..., and Cj vs Cj**

Continuing with the above example, which computed and plotted the model $y=x+2$, suppose we wanted to see the effect of changing the coefficient of x from 1 to 2; that is, $y = 2x+2$. First, use the LET command to compute the "new" y values:

MTB>**LET C3=2*C1+2**

Now use MPLOT to plot both models on the same graphs:

MTB>**MPLOt C2 C1, C3 C1**

The resulting plot is shown in Figure 2-1. Note that the first plot specified is given using the letter A, while the second uses B to mark the points. If two or more points are represented by the same point on the plot, a count of the number of points represented by the symbol is given, instead of the symbol itself.

```
MTB > MPLOT C2 C1,C3 C1
         C2
     24.0+
         -
         -
         -                                                                B
         -
     21.0+
         -
         -                                                        B
         -
         -
     18.0+
         -                                                 B
         -
         -                                          B
         -
     15.0+
         -                                   B
         -
         -
         -
     12.0+                            B                            A
         -                                                  A
         -                     B                      A
         -
      9.0+                                      A
         -
         -               B                A
         -                         A
      6.0+        B               A
         -
         -            A
         -     B     A
         -
      3.0+     A
         +---------+---------+---------+---------+---------+---------+C1
        0.0       2.0       4.0       6.0       8.0      10.0      12.0
```

FIGURE 2-1   A plot of the model y=2x+2 (denoted by B) and y=x+2
             (denoted by A).

In making plots of continuous functions using Minitab, two
issues are worth considering.   First, since the scatterplot
consists of a Cartesian point for each x value generated (until the
points are so close they can no longer be represented by a unique
symbol), the plot can be made to appear smoother by generating more
points.   (In the above example, we could use a step size of 0.5 to
make the plot appear more continuous.)   Second, better resolution
can be obtained by spreading the area to be plotted over as large a
space as possible.   The width and height of the plot can be con-
trolled in Minitab using the WIDTH and HEIGHT commands, which we
now address.

WIDTH AND HEIGHT

Minitab allows you to control the size of the plot using the HEIGHT and WIDTH commands. The format for WIDTH is:

MTB>**WIDTh** of plots = K spaces

where K can be varied in multiples of 10 from 30 to 100. The size of a space varies from printer to printer, with 10 spaces to an inch being typical. When Minitab is accessed, the WIDTH is set at 50 spaces until changed which, with a printer spacing of 10 per inch, gives a plot 5 inches wide.

It is important to note that Minitab adjusts the number of lines of height of the plot automatically to 1/2 the width setting. Thus the default width of 50 spaces gives 25 lines of height which, for a printer setting of 5 lines per inch, results in a vertical axis 5 inches high. For most printers the automatic height setting results in a set of axes that are approximately square. To set the height yourself, follow the WIDTH command with the HEIGHT command:

MTB>**HEIGht** of plot is K lines

where K can vary in multiples of 5 from 15 to 400. A setting of width 100 and height 50 typically gives a plot 7.5 by 10 inches.

The WIDTH and HEIGHT commands can also be combined on a single line:

MTB>**WIDTh** of plots is **K** spaces, height of plots is **K** lines

For example, the command (in abbreviated form)

MTB>**WIDT** 100 80

gives a plot 100 spaces wide and 80 lines high.

2.3 EXERCISES

1. Using the data provided in Exercises 2.2, plot the data generated by the model converting degrees Celsius to degrees Fahrenheit by plotting Fahrenheit temperatures versus Celsius temperatures.

2. Using the data you stored in Exercise 3 of Section 2.2:

   a. Plot y versus x.

   b. Change the coefficients of the model to K1=1, K2=6, and K3=9. Create a column of new y values, and plot the original model and new model on the same set of axes. Observe the changes in the model.

2.4   Example--An Inventory Problem: Minimizing The Cost Of Delivery And Storage

Consider a chain of gasoline stations. The manager wishes to determine how much and how often to deliver gasoline in order to minimize the average daily cost of maintaining the inventory while meeting the demand.   The inventory costs considered are the delivery costs and storage costs. Each time gasoline is delivered, the stations incur a charge of d dollars, which is in addition to the cost of the gasoline and is independent of the size of the order. It costs s dollars to store a gallon of gasoline one day. We define the following:

    s: storage cost per gallon per day
    d: delivery cost in dollars per delivery
    r: demand rate in gallons per day
    t: time in days between deliveries
    c: average daily cost

Under certain assumptions (see Giordano and Weir, op. cit. page 136, for a development of the model) the following model represents the average daily cost of delivering the gasoline every t days:

$$c = d/t + srt/2$$

The goal is to find a value of t that minimizes c for given values of d, s and r.   Each summand in the model has an interesting interpretation in terms of the physical problem so we would like to see how each summand behaves as a function of t. The first summand d/t represents the average daily cost of the delivery (the delivery cost "prorated" over the cycle length of t days).   The summand srt/2 represents the average daily cost of storing sufficient gasoline to meet demand if delivery occurs every t days.   For values of s = 0.10, d = 800 and r = 1000, let's graph the average delivery cost, average storage cost and average daily cost.   We begin by storing the three constants:

MTB>**LET K1=0.1**
MTB>**LET K2=800**
MTB>**LET K3=1000**

Next, generate values for t (the days) and give the column a name. We use t = 0.5 to achieve better resolution in the graph.

MTB>**SET C1**
DATA>**1:25/0.5**
DATA>**END**
MTB>**NAME C1 'TIME'**

In column C1 are values of t from 1 to 25 in increments of 0.5. We now compute the daily delivery cost d/t, store the values in C2, and give the column a name:

```
MTB>LET C2=K2/C1
MTB>NAME C2 'DELCOST'
```

In a similar manner we compute, store, and name the daily storage
cost srt/2:

```
MTB>LET C3=K1*K3*C1/2
MTB>NAME C3 'STORCOST'
```

Finally, we compute, store, and name the average daily cost c:

```
MTB>LET C4=C2+C3
MTB>NAME C4 'AVDLYCST'
```

Now we plot all graphs on the same axes:

```
MTB>MPLOT 'AVDLYCST' 'TIME','DELCOST' 'TIME', 'STORCOST' 'TIME'
```

Typically at this stage, we would experiment with the size of the
plots and ranges for the variables until we had a graph that we
liked. Note from the graph the existence of an "optimal" value
of t, giving a minimum average daily cost over the range of
values plotted. This optimal value of t is about 4 days, as seen
from the graph.

———————————————— EXHIBIT 2.4 ————————————————

Plotting The Average Daily Cost Of Delivering Gasoline

```
MTB > LET K1=0.1
MTB > LET K2=800
MTB > LET K3=1000
MTB > SET C1
DATA> 1:25/0.5
DATA> END
MTB > NAME C1 'TIME'
MTB > LET C2=K2/C1
MTB > NAME C2 'DELCOST'
MTB > LET C3=K1*K3*C1/2
MTB > NAME C3 'STORCOST'
MTB > LET C4=C2+C3
MTB > NAME C4 'AVDLYCST'
```

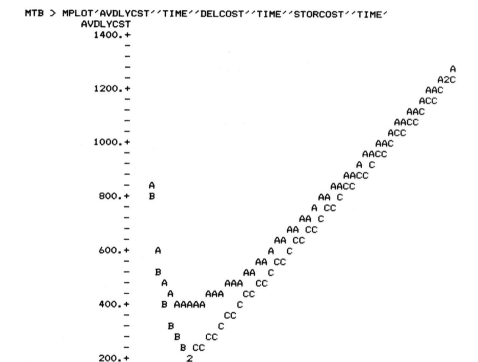

```
MTB > MPLOT'AVDLYCST''TIME''DELCOST''TIME''STORCOST''TIME'
        AVDLYCST
      1400.+
           -
           -
           -
           -                                                        A
      1200.+                                                       A2C
           -                                                      AAC
           -                                                     ACC
           -                                                    AAC
           -                                                   AACC
           -                                                  ACC
      1000.+                                               AAC
           -                                              AACC
           -                                            A C
           -                                           AACC
           -                                         AACC
       800.+    A                                  AA C
           -    B                                 A CC
           -                                    AA C
           -                                  AA CC
           -                                AA CC
       600.+    A                         A  C
           -                             AA CC
           -    B                      AA  C
           -     A                  AAA  CC
           -      A       AAA      CC
       400.+    B AAAAA        C
           -              CC
           -      B        C
           -       B    CC
           -        B CC
       200.+         2
           -        CC BBB
           -      C       BBBBB
           -    CC           BBBBBBBBBB
           -    C              BBBBBBBBBBBBBBBBBBBBBBBBBBB
         0.+
            +---------+---------+---------+---------+---------+---------+TIME
           0.0       5.0      10.0      15.0      20.0      25.0      30.0
```

3

MODELING USING PROPORTIONALITY

DATA ENTRY:	**SET**
	READ
ARITHMETIC COLUMN OPERATIONS:	
	ADD
	SUBTRACT
	MULTIPLY
	DIVIDE
	RAISE
	NATURAL
	LOGARITHM
	EXPONENTIAL
DATA CORRECTION:	
	SUBSTITUTE
	INSERT
	RECODE
	OMIT

Submodels are often constructed using proportionality arguments. Quite often large data sets are available which can be used to test graphically a proposed proportionality. If the test indicates that the assumed proportionality is reasonable, an initial estimate of the constant of proportionality can be made from the same graph. The procedure for testing a proportionality submodel often consists of the following steps:

1. Entering the data observed for the dependent and independent variables.

2. Plotting the raw data points to check smoothness and to identify potential data outliers.

3. Performing the transformations suggested by the submodel.

4. Plotting the transformed data to test the hypothesized proportionality.

5. Estimating the constant of proportionality.

In this chapter we study the Minitab commands necessary to perform the above operations. We begin by presenting several methods for entering discrete sets of observed data. In Chapter 2 you studied several techniques for transforming data; in this chapter you will learn several others. Then the process of making graphical plots worthy of checking the proposed proportionality, and for estimating the constants of proportionality, is presented. Finally, we apply the procedure to several modeling applications. Our goal in this chapter is to use Minitab to gain a visual, qualitative estimate of the worthiness of a submodel. In the next chapter we study analytical methods for fitting a type model and for determining how good it fits a given set of data.

3.1   Data Entry: Using The SET And READ Commands

In Chapter 2 we investigated the use of the SET command to enter patterned data into a column. Now we address the feature of entering any discrete data set.

SET

Probably the most convenient way to enter a column of observed data is to use the SET command as follows:

MTB>**SET** the following data in column **Cj**

Minitab will respond to this command with the prompt DATA>. The user then merely enters the data, separated by a space or comma, until all the data is entered. A carriage return can be used at any point to continue with the data entry: Minitab will again respond with DATA>. To terminate data entry and return to MTB> in the interactive mode it is good practice to indicate the completion of data entry by responding with an END to the DATA> prompt. Thus a typical session to enter a column of data would be:

```
MTB>SET C1
DATA>2 7 9
DATA>3.8 22
DATA>END
MTB>
```

Column 1 will now contain the numbers 2, 7, 9, 3.8, 22 in that order. It is a good idea to print the columns to verify the data entry. The SET command can also be used to enter data from an existing file into a specific column:

MTB>**SET 'filename'** into **Cj**

The filename must be enclosed in single quotes and cannot be the same as a column name.

Some General Rules For The SET Command

We list a few rules for the SET command:

1. Data must be separated by blank(s) or comma(s).

2. Data may be entered with or without a decimal point. Scientific notation is acceptable.

3. The missing data code * may be used. The * symbol indicates the value of the input variable is missing. If used, the entry will not be included in column operations such as "average,"

"standard deviation," and so forth (these are discussed later). Implications of the missing value code include:

       a. If one of the operands in an "arithmetic" or "function" command contains an element that is a * instead of a value, the result of the command is a * as well. For example, 2 + * = *.

       b. Most commands exclude from analysis all cases with a missing value in any input variable. For example AVERAGE computes the average of the numbers actually appearing in a column, ignoring any symbols * that may appear.

       c. The PLOT command plots only pairs (x,y) when both numbers are present. A count is given of all excluded points.

       d. A few commands do not work with the missing data code symbol *. In those cases a diagnostic is given.

READ

    If several columns are to be entered simultaneously, or an existing file is organized in rows, the user may prefer to use the READ command. Using this command the data is entered row by row:

MTB>**READ** the following data into Cj, Cj,..., Cj

In this command a carriage return signifies the end of a row and the END command signals the completion of the data entry. For example, consider the following sequence of commands:

MTB>**READ C1,C2**
DATA> 3   2
DATA> 1.4 3
DATA> 1   1.3
DATA> 9   5
DATA>**END**
MTB>

These commands enter the numbers 3, 1.4, 1, 9 into Column 1 and the numbers 2, 3, 1.3, 5 into Column 2. Note that if you enter a single column using the READ command, you can enter only one number at a time, followed by a <CR>.

    If you attempt to enter too many or too few numbers in a particular row, Minitab will warn you and you merely reenter the correct row. In the above case, if we had attempted to enter the third row as 1 1.3 9, Minitab would respond with

*TOO MANY VALUES-REENTER ROW

The READ command will accept a missing data symbol *, as does the SET command.

26

Some General Rules For The READ Command

A few rules for the READ command now follow:

1. The individual data values must be separated by blank(s) or comma(s).

2. Data lines may contain letters or words. However, any letters or words are ignored (i.e., they are not read as data).

3. Data can be entered with or without a decimal point, or in scientific notation.

4. In each row an entry must be made for each column identified in the command. (A missing data symbol * may be used.)

5. Since the carriage return <CR> indicates the end of a row, a continuation line (if one is needed to include all the data points) must be indicated by placing the symbol & in the last nonblank space at the end of the line.

As was the case in the SET command, the READ command may be used to enter data from an existing file ordered by rows:

MTB>READ 'filename' into columns Cj, Cj,..., Cj

Again, the filename must be enclosed in single quotes and the name cannot be the same as any column name.

A Minitab Session Illustrating Data Entry And Verification

Exhibit 3.1 illustrates entering, verifying, and naming data pertaining to the length and weight of bass caught during a fishing derby. In Section 3.4.1 we suggest several models for predicting the weight of a bass as a function of some readily measurable dimensions. In the Exhibit, notice that the data is entered in columns, the columns are named, and the data is printed to verify its correct entry. Finally, the data is saved in a file called "BASS" for use in Section 3.4.

Data Entry And Verification

```
MTB > READ C1 C2
DATA> 14.5 27
DATA> 12.5 17
DATA> 17.25 41
DATA> 14.5 26
DATA> 12.625 17
DATA> 17.75 49
DATA> 14.125 23
DATA> 12.625 16
DATA> END
     8 ROWS READ
MTB > NAME FOR C1 IS 'LENGTH',FOR C2 IS 'WEIGHT'
MTB > PRINT C1 C2
 ROW   LENGTH  WEIGHT

  1    14.500     27
  2    12.500     17
  3    17.250     41
  4    14.500     26
  5    12.625     17
  6    17.750     49
  7    14.125     23
  8    12.625     16

MTB > SAVE 'BASS'
```

Correcting Erroneous Entries

In the above illustrations using the READ and SET commands, we printed our data to verify that all elements had been entered correctly. But what happens if we find a mistake in an element or have omitted an element entirely? One possibility that comes to mind is to reuse the READ or SET command to reenter the column. If you have several mistakes, this procedure may very well be the best option. However, Minitab has several other options for correcting data, some of which include the use of the following commands: LET, SUBSTITUTE, INSERT, OMIT, and RECODE. We next discuss these commands in order.

LET

Individual elements of a column may be referenced by placing the row number of the element within parenthesis, following the column specification. Thus, the command

MTB>**LET C1(5)=28.3**

places the constant 28.3 in row 5 of column C1, replacing whatever value was previously there.

SUBSTITUTE

The SUBSTITUTE command has the following format:

MTB>**SUBS**titute the constant **K** into row **K** of column **Cj**

where the missing data symbol * may be substituted for the first K. Another format for the command is:

MTB>**SUBS**titute the number in row **K** of column **Cj**
　　　　　into row **K** of column **Cj**

Both forms of the SUBSTITUTE command replace one number in a column with another specified number, either directly (as in the first case) or indirectly (as in the second case). This command is now obsolete, but still acceptable. The preferred way of making corrections is to use the LET command to indicate the row in the particular column you wish to correct.

INSERT

If a data value has been omitted, the INSERT command is useful. Suppose we print C1 and C2 to obtain the following output:

ROW	C1	C2
1	61	67
2	96	131
3	0.5	0.8
4	14	13
5	65	64
6	115	123
7	12	0.5
8		17

Next we notice that the number 0.8 has been omitted between the 115 and 12 of Column C1. We can then use the INSERT command to make the adjustment:

MTB>**INSE**rt between rows 6 and 7 of **C1**
DATA>**0.8**
DATA>**END**
MTB>**PRINT C1 C2**

The output will now read as follows:

29

ROW	C1	C2
1	61	67
2	96	131
3	0.5	0.8
4	14	13
5	65	64
6	115	123
7	0.8	0.5
8	12	17

OMIT

Often we have an element of data that we wish to omit from a column. Perhaps it is an erroneous entry, or maybe we have decided it is an outlier that should be discarded. In such situations we can invoke the OMIT command. The format for the command is:

MTB>**OMIT K** from column **Cj** and put the remaining contents in **Cj**

The entire first column, less all elements having the value K, is placed in the second column. If desired, the first column can be maintained by designating a different second column.

Note that the OMIT command omits <u>all</u> occurrences of the value K. If you want to omit only some occurrences of the value K there are several alternative procedures (although Minitab does not have a specific command to accomplish this task). If there are only a few values to omit, you might consider using a combination of the LET and OMIT commands. For example, suppose the column C1 is as follows:

C1

1
2
1
3

We wish to omit only the second occurrence of the number 1; that is, the 1 appearing in Row 3. If we employ the command

MTB>**OMIT 1 C1**

both 1's will be discarded. Instead, consider the following combination of commands:

MTB>**LET C1(3)=99.9**
MTB>**OMIT 99.9 C1** put remaining contents in **C1**

Here the LET command replaces the 1 in row 3 with 99.9 (which is a unique value in C1). The OMIT command then discards the 99.9 so that C1 becomes as follows:

30

C1

1
2
3

The above technique is tedious if there are many values to omit, or if corresponding values in other columns need to be omitted as well. For example, you may decide to omit some data points, which are ordered pairs, because you consider them to be outliers. Thus, you wish to delete both an x, and its corresponding y, value. If you have more than one observation at that value of x, you may have a problem. Consider the following alternate format of the OMIT command:

MTB>**OMIT Ej** (unique value) from **Cj** together with the corresponding
rows in **Cj,...,Cj** and put into **Cj,...,Cj**

The above version of the OMIT command discards the identified value and the corresponding rows from all the designated columns. But we still face the problem of uniquely identifying the element we wish to discard. To accomplish this task, first generate an "index column" that indexes the rows. Then invoke the OMIT command. For example, suppose the following columns are given:

C4	C5
2	6
5	9
5	12
8	12

We have examined a scatterplot and decided that the data point (5,12) is an outlier, so we wish to discard it. However, there are two values of 5 in C4 and two values of 12 in C5. Thus, we begin by first creating an index column in an available column location:

MTB>**SET C3**
DATA>**1:4**
DATA>**END**

Our worksheet now contains the following data:

C3	C4	C5
1	2	6
2	5	9
3	5	12
4	8	12

Note that the elements in column C3 are unique, by design. We then omit the undesired element from C3 together with the corresponding elements in C4 and C5 using the OMIT command:

MTB>**OMIT 3 C3** and corresponding rows in **C4 C5** place in **C3-C5**

The above procedure yields the following worksheet:

C3	C4	C5
1	2	6
2	5	9
4	8	12

(We note that Minitab has a CHOOSE command with capabilities and formats analogous to the OMIT command, except that designated elements and corresponding elements in designated columns are retained and placed in specified columns.)

RECODE

Minitab allows us to change all entries of a certain value K to a new value, using the RECODE command. The format for the command is:

MTB>**RECO**de **K** in **Cj** to **K**, put into **Cj**

The above command changes all occurrences of the first value K to the second value K, and then puts the resulting new column in the second designated column. For example, suppose we had placed a missing value symbol * in C2 and then later obtained the value 3 for the missing value. The value 3 can be substituted for the symbol * using the RECODE command:

MTB>**RECO**de * **C2 3 C2**

The above command replaces column C2 with the updated column C2.

3.1 EXERCISES

1. Use the SET command to enter the following data sets. Print the columns and verify the entry.

    a.   x| 57.9 108.2 149.6 227.9 778.3 1427 2870 4497 5902

    b.   e|0 19 57 94 134 173 216 256 297 343 390
            s|5 10 20 30 40 50 60 70 80 90 100

2. Use the READ command to enter the following data sets. Verify the data entry.

    a.   x|0 1 2 3 4
           y|2 3 4 5 6

    b.   x|7 14 21 28 35 42
           y|8 41 133 250 280 297

c.

x	0	10	20	30	40	50
y	1900	1910	1929	1930	1940	1950
z	1.0	2.01	4.06	8.17	16.44	33.12

3. Use either the SET or READ commands to enter the following table. Verify the data entry.

A	149.3	135.2	126.6	107.6	103.7	113.4	127.8	142.5	153.5
B	107.3	100.9	98	88.9	86.6	91.5	97.3	107.6	111.5
C	26.8	17.9	12	3.3	2.1	4.3	13.6	15.2	22.5
D	13.5	15.2	15.9	15.1	14.9	17.3	17.3	20.3	19.6

4. Enter the following data set:

x	100	200	300	400	500	700
y	205	430	677	*	945	1872

a. Use the RECODE command to replace the symbol * with the value 945.

b. Use the LET command to replace the 945 corresponding to the x value of 500 with a new y value of 1233.

c. Insert the following data points into the set:

x	600	800
y	1542	2224

d. Print out x and y for verification.

3.2   Arithmetic Column Operations

The simplest proportionality argument would be $y \propto x$. To test the model we would put the observations for x in one column and the corresponding observations for y in a second column. A plot of y versus x should then approximate a straight line projected through the origin. More typically, a proportionality submodel involves a transformation of some sort, such as $y \propto x^3$ or $y \propto \ln x$. We could enter the observations for x and y, but then we would need to transform x before plotting y versus either x^3, or ln x, in the two cases cited. To accomplish these tasks, we could use the LET command and the functions presented in Chapter 2. However, Minitab allows an alternative procedure. In this section we discuss operations that can be performed directly on the columns. Among the many operations that are available are:

ADD, SUBTRACT, MULTIPLY, DIVIDE, RAISE, ABSOLUTE, SQRT, LOGTEN, LOGE, ANTILOG, EXPONENTIATE, ROUND, SIN, COS, TAN, ASIN, ACOS, ATAN

The command format is:

MTB>**CHOSEN ARITHMETIC COMMAND Ej** (to, from, or by) **Ej,...,Ej**, put into **Cj**

where E denotes either a number, stored constant, or a column.

Example 1 Summing Columns

ADD C1 TO C2, and put the result into column C3, or in abbreviated form:

MTB>**ADD C1 C2, C3**

The result is as follows:

C1	C2	C3
1.0	1.5	2.5
2.0	3.0	5.0
3.0	6.0	9.0

Example 2 Summing A Constant Into A Column

MTB>**ADD 3 C2 C4**

results in:

	C2	C4
3	1.5	4.5
3	3.0	6.0
3	6.0	9.0

The above operation could also be accomplished by first storing 3 as a constant:

MTB>**LET K1=3**
MTB>**ADD K1 C2 C4**

Example 3 Subtracting Columns

MTB>**SUBTRACT** C1 from C2, put into C5
MTB>**PRINt** C2 C1 C5

C2	C1	C5
1.5	1.0	.5
3.0	2.0	1.0
6.0	3.0	3.0

Example 4   Multiplying Columns

```
MTB>MULTIPLY C2 by C1, put into C6
MTB>PRINt C2 C1 C6

    C2          C1          C6

    1.5         1.0         1.5
    3.0         2.0         6.0
    6.0         3.0        18.0
```

Example 5   Dividing Columns

```
MTB>DIVIDE C2 by C1, put into C7
MTB>PRINt C2 C1 C7

    C2          C1          C7

    1.5         1.0         1.5
    3.0         2.0         1.5
    6.0         3.0         2.0
```

Example 6   Raising A Column To A Power

```
MTB>RAISE C2 to the power 2, put into C8
MTB>PRINt C2 C8

    C2          C8

    1.5          2.25
    3.0          9.0
    6.0         36.0
```

Summary Of Formats For The Arithmetic Column Operations

```
ADD Ej to Ej...to Ej, put into Cj
SUBTRACT Ej from Ej, put into Cj
MULTIPLY Ej by Ej...to Ej, put into Cj
DIVIDE Ej by Ej, put into Cj
RAISE Ej to the power Ej, put into Cj
ABSOLUTE value of Ej, put into Cj
SQRT of Ej, put into Cj
LOGTEN of Ej, put into Cj
LOGE of Ej, put into Cj
ANTILOG of Ej, put into Cj
EXPONENTIATE Ej, put into Cj
ROUND to nearest integer Ej, put into Cj
SIN  of Ej, put into Cj
COS  of Ej, put into Cj
TAN  of Ej, put into Cj
ASIN of Ej, put into Cj
ACOS of Ej, put into Cj
ATAN of Ej, put into Cj
```

where E denotes either a number, stored constant, or a column. It should be noted that the results can be put into one of the original columns if desired. For example, the command

MTB>**RAISE C2 2, C2**

squares the numbers in Column 2 and leaves them in Column 2.

3.2   EXERCISES

Perform the indicated arithmetic column operations for the following data sets.

1.   ADDITION

   a.   Show that y = x + 2 for the data given in Exercise 3.1.2a.

   b.   Find E = A + B + C + D

A	149.3	135.2	126.6	107.6	103.7	113.4	127.8	142.5	153.5
B	107.3	100.9	98	88.9	86.6	91.5	97.3	107.6	111.5
C	26.8	17.9	12	3.3	2.1	4.3	13.6	15.2	22.5
D	13.5	15.2	15.9	15.1	14.9	17.3	17.3	20.3	19.6
E									

2.   SUBTRACTION

   a.   Show that y - x = 2 for the data given in Exercise 3.1.2a.

b. Find F = B + C - D for the data given in Exercise 3.2.1b above.

3. MULTIPLICATION

For the data given in Exercise 3.2.1b above:

a. Find A*B and C*D.

b. Compute A * B + C * D.

c. Multiply D by 3.5.

4. DIVISION

For the data given in Exercise 3.2.1b above:

a. Find B/C.

b. Find (A * C)/(B * D).

5. THE RAISE-TO-A-POWER COMMAND

a. For the data given in Exercise 3.1.1a, find x^2, x^3, $x^{0.5}$.

b. For the data given in Exercise 3.1.1b, compute $e^2/2 + s^2/4$.

3.3 Testing Proportionality Arguments Using Minitab

Testing a proportionality argument often involves the following steps:

1. Entering the raw observed data.
2. Plotting the raw data to check for smoothness and outliers.
3. Making any necessary transformations.
4. Plotting the transformed data to test the proportionality.
5. Estimating the constant of proportionality.

Step 4 involves two crucial questions: Does the transformed data approximate a line? Does the line, when projected, pass through the origin? In this section we discuss and demonstrate the above five steps using Minitab.

In Step 4 it must first be decided if the transformed data actually lies approximately along a straight line. Two suggestions may be helpful. First, since Minitab rounds the data, you will want to use the HEIGHT and WIDTH commands to get as large a plot as possible. Second, for a given size plot, the data points should occupy a large portion of the space and not be all squeezed together. As discussed in Chapter 2, Minitab automatically scales the axes to spread the data over as large a space as possible.

37

The second part of Step 4 requires you to determine if the line, when projected, passes through the origin. This can readily be done by specifying that both the dependent and independent variables include the origin in the PLOT command.

To estimate the constant of proportionality from your graph (in Step 5), you want to estimate the slope of the best line you can fit graphically. Again, a large plot is advantageous and the origin does not necessarily need to be included. Thus you will often want to have _two_ plots of the transformed data: a large plot with the data spread out, and another one containing the origin. We now illustrate these ideas with several examples.

Example 1   Testing A Proportionality Model

In the following data ℓ represents length and V represents volume:

ℓ	0.6	1.0	2	4	7	20
V	0.1	0.7	6	100	210	4000

A modeler wishes to test the model $V \propto \ell^3$.

Step 1.   Enter the observed data and verify that the input is correct.

This is easily done in Minitab using the following commands:

```
MTB>SET V in C1
DATA> .1 .7 6 100 210 4000
DATA> END
MTB>SET L in C2
DATA> .6 1 2 4 7 20
DATA> END
MTB>NAME for C1 is 'VOLUME', for C2 is 'LENGTH'
MTB>PRINT C1 C2
```

Step 2.   Check for smoothness and identify potential outliers.

```
MTB>PLOT C1 versus C2
```

Step 3.   Make the transformations suggested by the data. Print the results to find the range of ℓ^3.

```
MTB>RAISE C2 to the power 3, put into C3
MTB>NAME for C3 is 'LNGTH**3'
MTB>PRINT C1 C2 C3
```

Step 4. Plot the transformed data to test for the following characteristics:

    a. A straight line:

MTB>**HEIGHT/WIDTH** (insert values if appropriate)
MTB>**PLOT C1 vs C3**

    b. Projection through the origin:

MTB>**PLOT C1 from 0 to 4000 vs C3 from 0 to 8000**

Step 5. Estimate the constant of proportionality. "Eyeball" a line onto the first graph plotted in Step 4; then pick two points (x_1, y_1) and (x_2, y_2) on or near this "best" line. The line segment between the chosen points should approximate the "eyeball" line you have drawn, and the points should be sufficiently far apart to avoid division by a small number. Now, the constant of proportionality is the slope

$$K1 = (y_2 - y_1)/(x_2 - x_1)$$

of the line segment between the chosen points. For the points identified in Exhibit 3.3.1 we have:

MTB>**LET K1=(4000-100)/(8000-64)**
MTB>**PRINT the value K1**

The constant slope K1 = .491431 is printed. Qualitatively we assume the proposed model is reasonable, and the graphical fit of $V = 0.5\ell^3$ is suggested.

Using the procedures discussed in Chapter 2, let's see how well the continuous model $V = 0.5\ell^3$ fits the set of observed data by plotting the model and overlaying the data set. The Minitab commands are:

MTB>**SET C4**
DATA>**1:20**
DATA>**END**
MTB>**LET C5=.5*C4**3**
MTB>**MPLOT C1 vs C2, C5 vs C4**

What do you conclude? The actual Minitab session and results for this example are presented in the following Exhibit 3.3.1.

———————————————— EXHIBIT 3.3.1 ————————————————

Testing A Proportionality Model

```
MTB > SET VOLUME IN C1
DATA> .1 .7 6 100 210 4000
DATA> END
MTB > SET LENGTH IN C2
DATA> .6 1 2 4 7 20
DATA> END
MTB > NAME FOR C1 IS 'VOLUME',FOR C2 IS 'LENGTH'
MTB > PRINT C1 C2
 ROW    VOLUME   LENGTH

   1       0.1      0.6
   2       0.7      1.0
   3       6.0      2.0
   4     100.0      4.0
   5     210.0      7.0
   6    4000.0     20.0

MTB >


MTB > PLOT C1 C2
          VOLUME
          4500.+
              -
              -                                              *
              -
              -
              -
          3000.+
              -
              -
              -
              -
          1500.+
              -
              -
              -
              -
              -
           0.+  ** *    *          *
             +---------+---------+---------+---------+---------+LENGTH
            0.0       5.0      10.0      15.0      20.0      25.0
```

```
MTB > RAISE C2 TO THE 3 POWER,PUT INTO C3
MTB > NAME FOR C3 IS 'LNGTH**3'
MTB > PRINT C1 C2 C3
 ROW   VOLUME   LENGTH   LNGTH**3

   1      0.1      0.6       0.22
   2      0.7      1.0       1.00
   3      6.0      2.0       8.00
   4    100.0      4.0      64.00
   5    210.0      7.0     343.00
   6   4000.0     20.0    7999.98

MTB >
```

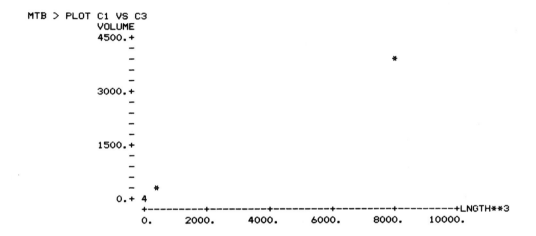

```
MTB > NOTE ESTIMATING THE SLOPE.
MTB > LET K1=(4000-100)/(8000-64)
MTB > PRINT K1
K1       0.491431
MTB > SET C4
DATA> 1:20
DATA> END
MTB > LET C5=.5*C4**3
MTB >
```

```
MTB > WIDTH 100
MTB > MPLOT C1 VS C2,C5 VS C4
```

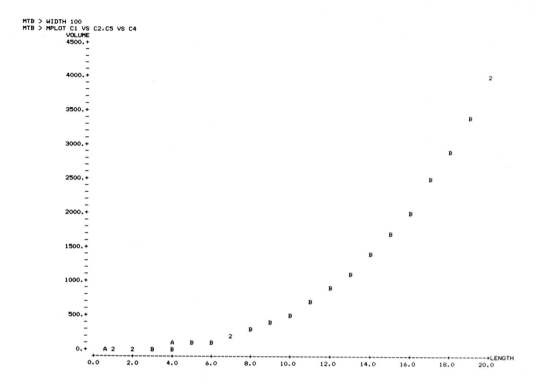

Example 2   Rejecting A Poor Model

This sample demonstrates that the qualitative information
obtained from a Minitab plot can be sufficient to reject a poor
model.   In the following data set, P represents the population of
fruit flies and t represents the time spent (in days) in incubation.   The modeler suspects that the population of flies is proportional to the time spent in incubation: $P \propto t$.

t	7	14	21	28	35	42
P	8	41	133	250	280	297

We test the proposed proportionality using Minitab:

```
MTB>SET P in C1
DATA> 8 41 133 250 280 297
DATA>END
MTB>SET T in C2
DATA> 7 14 21 28 35 42
DATA>END
MTB>NAME for C1 is 'FLIES', for C2 is 'TIME'
MTB>PRINT C1 C2
```

```
MTB>PLOT C1 C2
MTB>PLOT C1 from 0 to 42 vs C2 from 0 to 300
```

The actual Minitab session and output is presented in Exhibit
3.3.2. Note that a line through the origin does not approximate
the data reasonably well.   Thus, we reject the proportionality
assumption P ∝ t and return to the modeling process to investigate other possible relationships.

───────────────────────── EXHIBIT 3.3.2 ─────────────────────

A Poor Proportionality Model

```
MTB > SET FLY POPULATION IN C1
DATA> 8 41 133 250 280 297
DATA> END
MTB > SET TIME IN C2
DATA> 7 14 21 28 35 42
DATA> END
MTB > NAME FOR C1 IS 'FLIES', C2 IS 'TIME'
MTB > PRINT C1 C2
 ROW   FLIES    TIME

   1       8       7
   2      41      14
   3     133      21
   4     250      28
   5     280      35
   6     297      42

MTB >
```

```
MTB > PLOT C1 VS C2
          FLIES
         300.+                                              *
             -
             -
             -
             -
         240.+                                        *
             -
             -                              *
             -
         180.+
             -
             -
             -                     *
         120.+
             -
             -
             -
          60.+
             -
             -               *
             -
             -       *
           0.+
             +---------+---------+---------+---------+---------+---------+--TIME
            0.0       8.0      16.0      24.0      32.0      40.0      48.0

MTB > SAVE 'TIMEFLYS'
```

Example 3   Checking If A Line Projects Through The Origin

A modeler suspects the model $y \propto \ln x$ and has gathered the following data:

x	8.1	22.1	60.1	165
y	1	2	3	4

and uses Minitab to test the proportionality:

```
MTB>SET X in C1
DATA> 8.1 22.1 60.1 165
DATA>END
MTB>SET Y in C2
DATA> 1 2 3 4
DATA>END
MTB>NAME C1 'x' C2 'y'
MTB>PRINt C1 C2
MTB>LOGE C1 put into C3
MTB>NAME C3 'LN X'
MTB>PRINt C1 C2 C3
MTB>PLOT C2 vs C3
```

Examination of the output in Exhibit 3.3.3 reveals that the data do lie approximately along a straight line. However, we need to ensure that the line projects through the origin:

```
MTB>PLOT C2 0 4 C3 0 6
```

Notice from the plot that the line fails to project through the origin; thus we reject the proposed proportionality. Also note that if we had not checked whether the line passes through the origin, we might have used the first graph and determined a constant of proportionality (which would have been incorrect).

EXHIBIT 3.3.3

Testing A Proportionality: Does The Line Project Through The Origin?

```
MTB > SET X IN C1
DATA> 8.1 22.1 60.1 165
DATA> END
MTB > SET Y IN C2
DATA> 1 2 3 4
DATA> END
MTB > NAME C1 'X',C2 'Y'
MTB > PRINT C1 C2
 ROW       X     Y

    1      8.1    1
    2     22.1    2
    3     60.1    3
    4    165.0    4

MTB >


MTB > LOGE C1 PUT INTO C3
MTB > NAME C3 'LN X'
MTB > PRINT C1-C3
 ROW       X     Y      LN X

    1      8.1    1    2.09186
    2     22.1    2    3.09558
    3     60.1    3    4.09601
    4    165.0    4    5.10594

MTB >
```

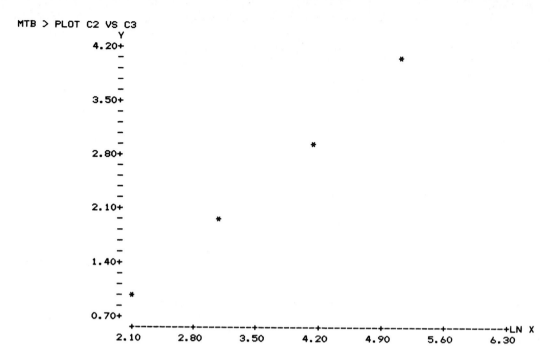

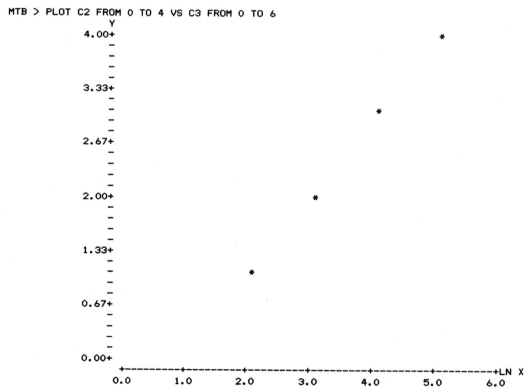

3.4   Examples Using Minitab To Test Proportionality Submodels

In this section we present several examples illustrating the use of the Minitab commands presented thus far in modeling applications.

Example 1   A Bass Fishing Derby

Consider a sport fishing club that for conservation purposes wishes to encourage its membership to release their fish immediately after catching them. On the other hand, the club wishes to make awards based on the total weight of fish that are caught. You might suggest that each individual carry a small portable scale. However, portable weight scales tend to be inconvenient and inaccurate, especially for smaller fish. Thus we define our problem as follows:

<u>Predict the weight of a fish in terms of some easily measurable dimensions.</u>

If we let W represent the weight of a fish in ounces, ℓ represent the length of a fish in inches, and g its girth (circumference of the fish at its widest point) in inches, we can suggest the following models (see Giordano and Weir, <u>op. cit.</u>, page 57 for a development):

1. $W \propto \ell^3$

2. $W \propto g^3$

3. $W \propto g^2 \ell$

4. $W \propto g \ell^2$

We have collected the following set of data and wish to test each of the above models:

Length, ℓ(in.)	14.5	12.5	17.25	14.5	12.625	17.75	14.125	12.625
Girth, g(in.)	9.75	8.375	11.0	9.75	8.5	12.5	9.0	8.5
Weight, W(oz.)	27	17	41	26	17	49	23	16

In Exhibit 3.4.1, we first retrieve the 'BASS' data file from a previous session, then we enter the above data set, make the transformations suggested by the model, plot the transformed data, determine whether each transformed plot approximates a line through the origin, and estimate the various constants of proportionality, if appropriate. Graphical estimates of the constants of proportionality yield the following models:

1. $W = .0083 \ \ell^3$

2. $W = .028 \ g^3$

3. $W = .0183 \, g^2 \ell$

4. $W = .0125 \, g\ell^2$

Which model fits the data better? In Chapter 4 we fit and analyze the above models analytically to help answer the question "Which model is best?"

───────────────────────── EXHIBIT 3.4.1 ─────────────────────────

Proportionality Models For The Bass Fishing Derby

```
MTB > RETR 'BASS'
MTB > RAISE C1 TO 3 POWER PUT INTO C3
MTB > NAME C3 'LNGTH**3'
MTB > SET GIRTH INTO C4
DATA> 9.75 8.375 11 9.75 8.5 12.5 9 8.5
DATA> END
MTB > NAME C4 'GIRTH'
MTB > RAISE C1 TO 2 POWER PUT INTO C5
MTB > NAME C5 'LNGTH**2'
MTB > RAISE C4 TO 2 POWER PUT INTO C6
MTB > RAISE C4 TO 3 POWER PUT INTO C7
MTB > NAME C6 'GIRTH**2',C7 'GIRTH**3'
MTB > MULTIPLY C1 BY C6,PUT INTO C8
MTB > NAME C8 'L G**2'
MTB > MULTIPLY C4 BY C5,PUT INTB C9
MTB > NAME C9 'G L**2'
MTB > PRINT C1 C2 C3 C7 C8 C9
```

ROW	LENGTH	WEIGHT	LNGTH**3	GIRTH**3	L G**2	G L**2
1	14.500	27	3048.61	926.86	1378.40	2049.93
2	12.500	17	1953.12	587.43	876.76	1308.59
3	17.250	41	5132.93	1331.00	2087.25	3273.18
4	14.500	26	3048.61	926.86	1378.40	2049.93
5	12.625	17	2012.30	614.12	912.15	1354.82
6	17.750	49	5592.35	1953.12	2773.43	3938.27
7	14.125	23	2818.15	729.00	1144.12	1795.64
8	12.625	16	2012.30	614.12	912.15	1354.82

```
MTB > WIDTH 70
```

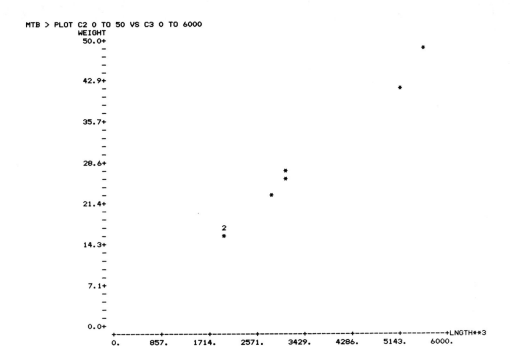

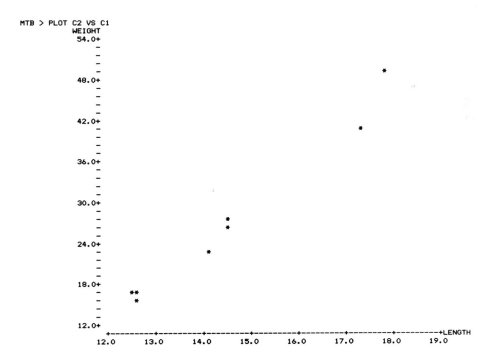

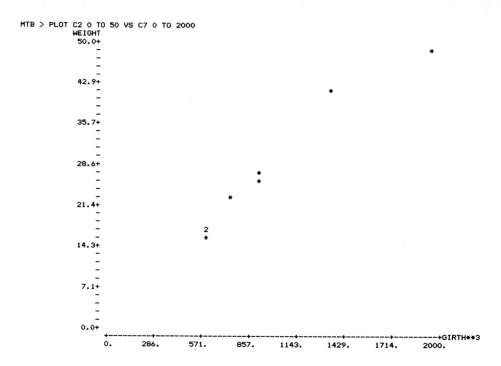

```
MTB > PLOT C2 0 TO 50 VS C7 0 TO 2000
        WEIGHT
   50.0+
       -                                                                    *
       -
       -
       -
   42.9+
       -                                                        *
       -
       -
   35.7+
       -
       -
       -
   28.6+
       -                                   *
       -                                   *
       -
   21.4+                            *
       -
       -
       -                      2
       -                      *
   14.3+
       -
       -
       -
    7.1+
       -
       -
    0.0+
       +---------+---------+---------+---------+---------+---------+---------+GIRTH**3
      0.       286.      571.      857.     1143.     1429.     1714.    2000.
```

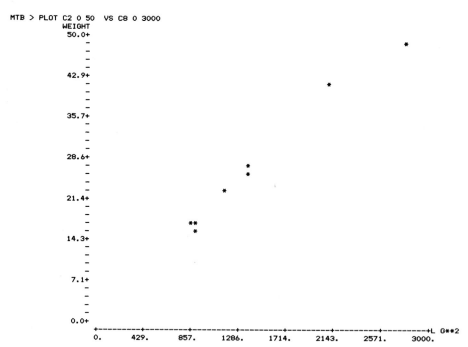

```
MTB > PLOT C2 0 50   VS C8 0 3000
        WEIGHT
   50.0+
       -                                                              *
       -
       -
   42.9+
       -                                                      *
       -
       -
   35.7+
       -
       -
       -
   28.6+
       -                                    *
       -                                    *
       -
   21.4+                              *
       -
       -                       **
       -                       *
   14.3+
       -
       -
    7.1+
       -
       -
    0.0+
       +---------+---------+---------+---------+---------+---------+---------+L G**2
      0.       429.      857.     1286.     1714.     2143.     2571.    3000.
```

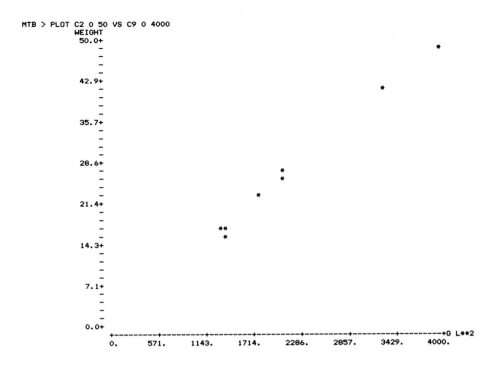

```
MTB > PLOT C2 0 50 VS C9 0 4000
        WEIGHT
    50.0+                                                    *
        -
        -
        -
    42.9+                                          *
        -
        -
        -
    35.7+
        -
        -
        -
    28.6+                              *
        -                              *
        -
        -                    *
    21.4+
        -
        -
        -           **
        -           *
    14.3+
        -
        -
        -
    7.1+
        -
        -
        -
    0.0+
        +---------+---------+---------+---------+---------+---------+---------+G L**2
        0.      571.     1143.     1714.     2286.     2857.     3429.    4000.
```

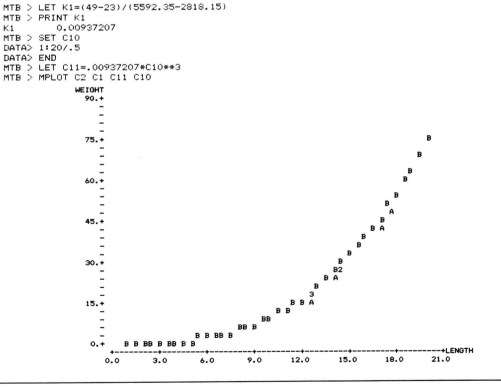

```
MTB > LET K1=(49-23)/(5592.35-2818.15)
MTB > PRINT K1
K1      0.00937207
MTB > SET C10
DATA> 1:20/.5
DATA> END
MTB > LET C11=.00937207*C10**3
MTB > MPLOT C2 C1 C11 C10
        WEIGHT
    90.+
        -
        -
        -
    75.+                                                          B
        -                                                       B
        -                                                    B
    60.+                                                 B
        -                                            B
        -                                          B
        -                                        B A
    45.+                                      B
        -                                   B A
        -                                 B
        -                              B
    30.+                           B
        -                        B2
        -                      B A
        -                    B
    15.+               B B A
        -            B B
        -          BB
        -       BB B
    0.+   B B BB B BB B B
        +---------+---------+---------+---------+---------+---------+---------+LENGTH
        0.0      3.0       6.0       9.0      12.0      15.0      18.0      21.0
```

Example 2   Vehicular Braking Distance

A popular "Rule of Thumb" often given to students in driver education classes is the "Two Second Rule" to prescribe a safe following distance. The rule states that if you stay two seconds behind the car in front, you have the correct distance no matter what your speed. Since the amount of time is constant (2 seconds), the rule suggests a proportionality between stopping distance and speed v. To test this rule we pose the following problem:

<u>Predict a vehicle's total stopping distance as a function of its speed.</u>

In the development of this model (see Giordano and Weir, <u>op.cit.</u>, page 64), total stopping distance is calculated as the sum of the reaction distance d_r and braking distance d_b. The following submodels are hypothesized in that development:

1. $d_r \propto v$

2. $d_b \propto v^2$

These submodels suggest that the total stopping distance d is given by the equation

$$d = k_1v + k_2v^2$$

The submodels are tested and calibrated using the following data set:

v	20	25	30	35	40	45	50	55	60	65	70	75	80
d_r	22	28	33	39	44	50	55	61	66	72	77	83	88
d_b	20	28	40.5	52.5	72	92.5	118	148.5	182	220.5	266	318	376
d	42	56	73.5	91.5	116	142.5	173	209.5	248	292.5	343	401	464

In Exhibit 3.4.2, each of the two submodels is tested by entering the appropriate data, making the suggested transformations, plotting the transformed data, and determining the constants of proportionality. Estimating these constants graphically, we obtain the following model:

$$d = 1.1v + 0.054v^2$$

Estimating Proportionality Submodels For
Vehicular Braking Distance

```
MTB > SET SPEED IN C1
DATA> 20:80/5
DATA> END
MTB > SET REACTION DISTANCE IN C2
DATA> 22 28 33 39 44 50 55 61 66 72 77 83 88
DATA> END
MTB > SET BRAKING DISTANCE IN C3
DATA> 20 28 40.5 52.5 72 92.5 118 148.5 182 220.5 266 318 376
DATA> END
MTB > LET C4=C2+C3
MTB > NAME C1 'SPEED',C2 'R DIST',C3 'B DIST', C4 'T DIST'
MTB > PRINT C1-C4
```

ROW	SPEED	R DIST	B DIST	T DIST
1	20	22	20.0	42.0
2	25	28	28.0	56.0
3	30	33	40.5	73.5
4	35	39	52.5	91.5
5	40	44	72.0	116.0
6	45	50	92.5	142.5
7	50	55	118.0	173.0
8	55	61	148.5	209.5
9	60	66	182.0	248.0
10	65	72	220.5	292.5
11	70	77	266.0	343.0
12	75	83	318.0	401.0
13	80	88	376.0	464.0

```
MTB > RAISE C1 2 C5
MTB > NAME C5 'SPEED**2'
MTB > PRINT C1 C5 C2 C3 C4
```

ROW	SPEED	SPEED**2	R DIST	B DIST	T DIST
1	20	400.0	22	20.0	42.0
2	25	625.0	28	28.0	56.0
3	30	900.0	33	40.5	73.5
4	35	1225.0	39	52.5	91.5
5	40	1600.0	44	72.0	116.0
6	45	2025.0	50	92.5	142.5
7	50	2500.0	55	118.0	173.0
8	55	3025.0	61	148.5	209.5
9	60	3600.0	66	182.0	248.0
10	65	4225.0	72	220.5	292.5
11	70	4900.0	77	266.0	343.0
12	75	5625.0	83	318.0	401.0
13	80	6400.0	88	376.0	464.0

```
MTB >
```

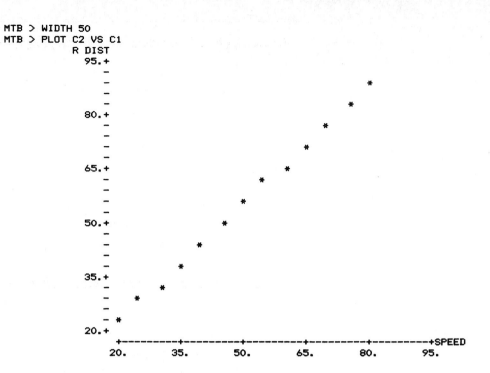

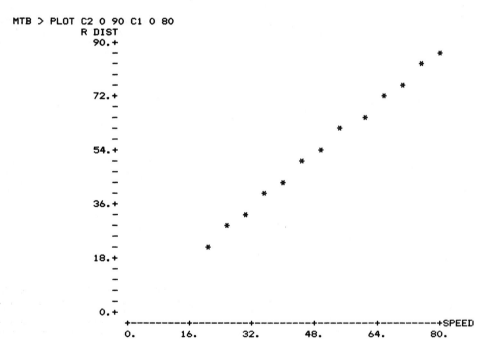

54

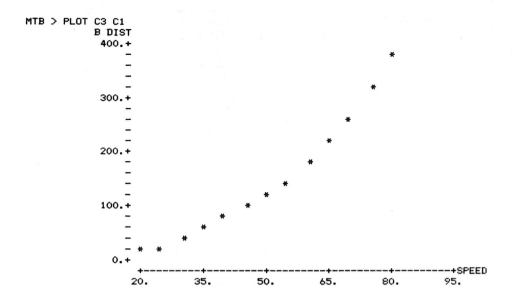

```
MTB > PLOT C3 C1
      B DIST
   400.+                                          *
       -
       -
       -
   300.+                                      *
       -
       -                                   *
       -                                *
   200.+                             *
       -                         *
       -                      *
       -                   *
   100.+                *
       -             *
       -          *
       -       *
     0.+  *  *
       +---------+---------+---------+---------+---------+SPEED
         20.      35.       50.       65.       80.       95.
```

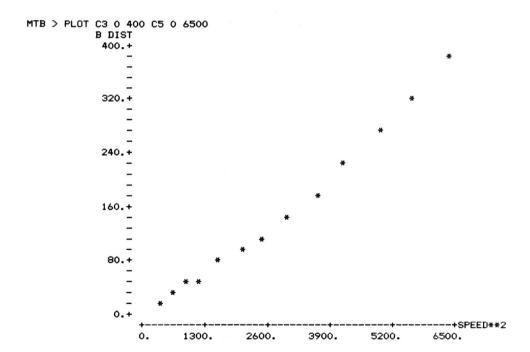

```
MTB > PLOT C3 0 400 C5 0 6500
      B DIST
   400.+                                                *
       -
       -
       -
   320.+                                           *
       -
       -                                       *
       -
   240.+                                  *
       -
       -                            *
       -
   160.+                      *
       -
       -                 *
       -              *
    80.+          *
       -
       -      *  *
       -    *
       -  *
     0.+
       +---------+---------+---------+---------+---------+SPEED**2
         0.      1300.     2600.     3900.     5200.     6500.
```

55

```
MTB > PLOT C4 C1
          T DIST
        500.+
            -
            -                                              *
            -
            -
        400.+                                          *
            -
            -
            -                                       *
            -
        300.+                                   *
            -
            -
            -                               *
            -
        200.+                          *
            -                       *
            -                    *
            -                 *
        100.+              *
            -           *
            -        *
            -     *
            - *
          0.+
            +---------+---------+---------+---------+---------+SPEED
           20.       35.       50.       65.       80.       95.


MTB > LET K1=(88-22)/(80-20)
MTB > PRINT K1
K1        1.10000
MTB > LET K2=(318-28)/(5625-625)
MTB > PRINT K2
K2        0.0580000
MTB > NOTE THE SLOPES ARE 1.1 AND .058   .
MTB >
```

```
MTB > SET C6
DATA> 20:80
DATA> END
MTB > LET C7=1.1*C6+.058*C6**2
MTB > NAME C6 'G SPEED',C7 'P DIST'
MTB > PLOT C7 C6
```

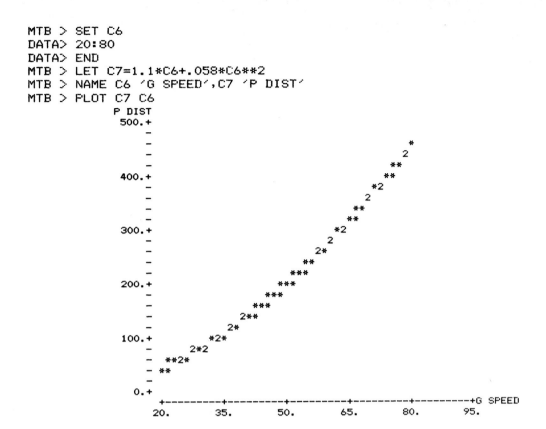

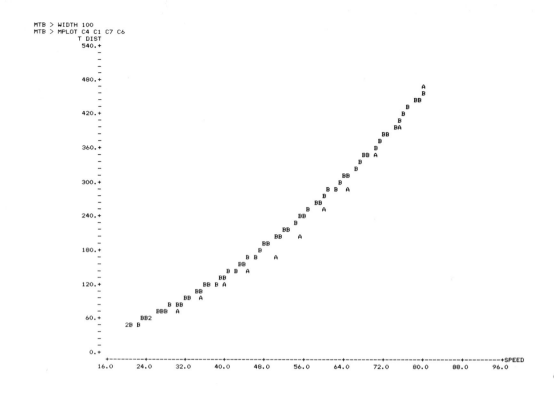

```
MTB > WIDTH 100
MTB > MPLOT C4 C1 C7 C6
        T DIST
       540.+
          -
          -
          -
          -
       480.+                                                                  A
          -                                                                  B
          -                                                                 BB
          -                                                              B
       420.+                                                            B
          -                                                            B
          -                                                           BA
          -                                                        BB
       360.+                                                      B
          -                                                   BB A
          -                                                 B
          -                                                B
       300.+                                            BB
          -                                            B
          -                                        B B A
          -                                       B
          -                                     BB
       240.+                                  BB
          -                                 B
          -                              BB
          -                             BB  A
       180.+                          B
          -                        B B    A
          -                       BB
          -                    B B  A
          -                   BB
       120.+                BB B A
          -               BB
          -             BB  A
          -           B BB
          -         BBB  A
        60.+      BB2
          -     2B B
          -
          -
         0.+
          +--------+--------+--------+--------+--------+--------+--------+--------+--------+--------+SPEED
        16.0     24.0     32.0     40.0     48.0     56.0     64.0     72.0     80.0     88.0     96.0
```

Example 3   The Elapsed Time Of A Tape Recorder

Most stereo cassette decks, reel-to-reel audiotape, and video cassette recording devices are equipped with a numerical counter. This tape counter provides a numerical reference point which can be used to index a recorded cassette or tape for playback purposes. The counter is incremented as the tape advances over the read/write heads. In many instances it is desired to relate the number displayed by the counter to time; for example, to find the playing time of a selection or to determine its tape count. Does the counter on your tape recorder operate in such a manner that the elapsed time is proportional to the displayed count? For example, suppose during the playing of a tape requiring 90 minutes to play, your counter goes from 0 to 1600. Does it take 45 minutes for the counter to reach 800? When your counter reaches 1400, does enough playing time still remain to record another album or particular selection? Thus we identify the following problem:

For a particular cassette deck or tape recorder equipped with a tape counter, relate the counter to the amount of playing time that has elapsed.

Letting t represent the elapsed playing time and c the counter reading, the following model is developed (see Giordano and Weir, op. cit., page 74):

$$t = k_2c + k_3c^2$$

One of the submodels requires testing the proportionality $c \propto \theta/2\pi$ where $\theta/2\pi$ is the number of revolutions of the take-up reel. In Exhibit 3.4.3 we use Minitab to test the submodel against the following data set:

c	100	200	300	400	500	600	700	800	900	1000	1100	1200	1300	1400
$\theta/2\pi$	110	221	330	439	543	654	762	869	976	1086	1200	1307	1422	1535

The model $t = k_2c + k_3c^2$ will be fit and tested in Chapter 4.

─────────────── EXHIBIT 3.4.3 ───────────────

**Testing The Proportionality Of The Counter Reading
vs The Number Of Revolutions Of The Take-up Reel**

```
MTB > SET REVOLUTIONS IN C1
DATA> 100:1400/100
DATA> END
MTB > SET COUNTER READINGS IN C2
DATA> 110 221 330 439 543 654 762 869 976 1086 1200 1307 1422 1535
DATA> END
MTB > NAME C1 'REVS',C2 'COUNTER'
MTB > PRINT C1 C2
 ROW    REVS   COUNTER

  1     100       110
  2     200       221
  3     300       330
  4     400       439
  5     500       543
  6     600       654
  7     700       762
  8     800       869
  9     900       976
 10    1000      1086
 11    1100      1200
 12    1200      1307
 13    1300      1422
 14    1400      1535

MTB >
```

```
MTB > WIDTH 60
MTB > PLOT C2 C1
        COUNTER
     1800.+
          -
          -
          -
          -                                                      *
     1500.+
          -                                                  *
          -
          -                                              *
          -
     1200.+
          -                                          *
          -
          -                                      *
          -
      900.+
          -                                  *
          -                              *
          -
          -                          *
      600.+
          -                      *
          -
          -                  *
          -              *
      300.+
          -          *
          -
          -      *
          -
        0.+
          +---------+---------+---------+---------+---------+---------+REVS
          0.       300.      600.      900.     1200.     1500.     1800.
```

3.4 PROJECTS

For Projects 1 through 7 below, complete the following requirements:

A. Develop and test a proportionality model for the problem identified.

B. Obtain a scatterplot of the given data.

C. Making the necessary transformations, test your proposed proportionality. Is the model reasonable?

D. Estimate the parameters of your model from the graph.

E. Use MPLOT to graph the proposed model with the data set superimposed on the same graph.

F. What conclusions do you draw?

60

1. Lumbercutters: Lumbercutters wish to use a readily available measurement to estimate the number of board feet of lumber in a tree. Let's assume they measure the diameter of the tree in inches at waist height. Develop a model that predicts board feet as a function of diameter in inches. The following data is provided for your test:

x	17	19	20	23	25	28	32	38	39	41
y	19	25	32	57	71	113	123	252	259	294

The variable x is the diameter of a ponderosa pine in inches and y is the number of board feet divided by 10.

2. Heart-rate in Birds: Warm-blooded animals use large quantities of energy to maintain body temperature because of heat loss through the body surface. In fact, biologists believe that the primary energy drain on a resting warm-blooded animal is the maintenance of body temperature.

   a. Construct a model relating blood flow through the heart to body weight. Assume that the energy available is proportional to the blood flow through the lungs, which is the source of oxygen. Assuming the least amount of blood needed is circulated, the amount of available energy will equal the amount used.

   b. The following data relate the weights of some birds to their heart rate in beats per minute. Construct a model that relates heart rate to body weight. Discuss the assumptions of your model. Use the data to check your model.

Bird	Body weight (grams)	Pulse rate
Canary	20	1000
Pigeon	300	185
Crow	341	378
Buzzard	658	300
Hen	2000	312
Domestic duck	2300	240
Wild duck	1100	190
Turkey	8750	193
Ostrich	71000	60-70

Data from A. J. Clark, Comparative Physiology of the Heart (New York, Macmillan, 1927), page 99.

3. Heart rate in Mammals: The following data relate the weights of some mammals to their heart rates in beats per minute. As in Problem 2, construct a model that relates heart rate to body weight. Discuss the assumptions of your model. Use the following data to check your modeling.

Animal	Body weight (grams)	Pulse rate
Vesperugo pipistrellas	4	660
Mouse	25	670
Rat	200	420
Guinea pig	300	300
Rabbit	2,000	205
Hare	3,000	64
Dog (a)	5,000	120
Dog (b)	30,000	85
Sheep	50,000	70
Man	70,000	72
Horse	450,000	38
Ox	500,000	40
Elephant	3,000,000	48

Data from A.J. Clark, Comparative Physiology of the Heart (New York, Macmillan, 1927), page 99.

4. Superstars: In the television show "Superstars" the top athletes from various sports compete against one another in a variety of events. Of course, the athletes vary considerably in height and weight. To compensate for this in the weightlifting competition, the body weight of the athlete is subtracted from his lift. What kind of a relationship between power and body weight does this suggest? Use the data provided below to check this assumption. The data displays the winning lifts for weight lifting at the 1976 Montreal Olympic Games.

Bodyweight class		Total winning lift (lb)		
	Max. weight (lb)	Snatch	Jerk	Total weight
Flyweight	114.5	231.5	303.1	534.6
Bantamweight	123.5	259.0	319.7	578.7
Featherweight	132.5	275.6	352.7	628.3
Lightweight	149.0	297.6	380.3	677.9
Middleweight	165.5	319.7	418.9	738.5
Light-heavyweight	182.0	358.3	446.4	804.7
Middle-heavyweight	198.5	374.8	468.5	843.3
Heavyweight	242.5	385.8	496.0	881.8

a. Physiological arguments have been proposed suggesting that the strength of a muscle is proportional to its cross-sectional area. Using this submodel for strength, construct a model relating lifting ability and body weight, assuming that the weight lifters are geometrically similar. List any additional assumptions you have made. Test your model against the data provided above.

b. Now consider a refinement to the above model. Suppose there is a certain amount of body weight which is independent of size in adults. Suggest a model which incorporates this refinement and test it against the data.

c. Criticize the use of the above data. What data would you really like to have in order to handicap the weightlifters? Who is the best weightlifter according to the work you have performed? Suggest a "Rule of Thumb" for determining the overall winning weightlifter in the "Superstars" show.

5. Heart weights: Are the hearts of animals geometrically similar? Use the data provided below to relate heart weight of an animal to some characteristic dimension.

Animal	Heart weight (grams)	Length of cavity of left ventricle
Mouse	0.13	0.55
Rat	0.64	1.0
Rabbit	5.8	2.2
Dog	102	4.0
Sheep	210	6.5
Ox	2030	12.0
Horse	3900	16.0

Data from A. J. Clark, Comparative Physiology of the Heart (New York, Macmillan, 1927), page 84.

6. Tape Recorders: Suppose the elapsed time t of a tape recorder is related to the counter reading c by the expression

$$t = (\pi/ms)[2r_0 c + (h_{eff}/m)c^2]$$

where
- s: linear speed of the tape recorder in playback mode
- m: constant of proportionality for $c = m(\theta/2\pi)$
- c: counter reading
- r_0: radius of the hub of the reel driving the counter
- h_{eff}: effective thickness of tape

A modeler suspects that a proportionality relationship exists between the counter reading and the number of revolutions of the reel driving the counter; that is, $c \propto \theta/2\pi$. The following data have been collected for a cassette tape recorder:

counter reading	30	60	90	120	150	180	210	240	270	300
revs, supply drive	26	57	92	130	171	216	266	321	383	456
revs, take-up drive	50	100	151	201	251	300	349	401	451	501

a. Use Minitab to determine whether the supply or the take-up drive operates the counter. For the drive chosen, determine the constant of proportionality for the relation $c = m(\theta/2\pi)$.

b. Given the following parameters:

linear speed:	1.75 inches per second
tape thickness:	0.3/500 inches
hub diameter:	$2r_0 = 0.85$ inches

Use the model suggested above to predict the playing time as a function of the counter reading and compare your answers with the actual data given below. Use the Minitab LET command to perform your calculations.

counter reading	30	60	90	120	150	180	210	240	270	300
elapsed time (sec)	75	156	245	335	432	533	640	754	871	995

For each counter reading display ERROR and PERCENT ERROR.

7. <u>Bass Fishing Derby</u>: Using the following data, verify that the weight W of a bass is proportional to the cube of a characteristic dimension ℓ ; i.e., $W \propto \ell^3$. Graphically estimate the constant of proportionality.

Length (in.)	Weight (lb-oz)	Length (in.)	Weight (lb-oz)	Length (in.)	Weight (lb-oz)
18	3-0	13	1-2	14 1/2	1-6
16 1/4	2-8	13 1/2	1-4	13 1/8	1-2
12 5/8	1-0	12 1/4	0-15	12 3/4	1-3
13 3/4	1-4	13 1/2	1-3	12	0-14
13 3/4	1-5	13 3/4	1-4	14	1-8
14 3/4	1-8	13 1/8	1-1	14 3/4	1-12
13 5/8	1-6	13 1/2	1-4	12 5/8	1-0
13 1/4	1-3	14 3/8	1-7	12 1/2	0-14
12 5/8	1-1	12 1/2	1-0	15 5/8	2-1
13 1/4	1-3	13 3/4	1-4	13 1/2	1-4
12 1/4	1-1	13 1/2	1-5	13 7/8	1-6
13 1/2	1-8	12	0-13	13	1-5
13 1/8	1-4	13 1/4	1-4	13 5/8	1-0
15 3/4	2-4	12 1/4	0-13	14	1-7
14 3/8	1-10	17	2-9	12 3/8	1-0
17	2-14	14 7/8	1-8	13 1/2	1-7
17 3/8	2-9	14 1/2	1-13	12 7/8	1-7
13 1/2	1-8	17	2-12	15 3/8	1-11
12 3/4	1-1	13 1/8	1-2	12 3/4	1-1
14 1/4	1-9	13 1/4	1-4	12 3/4	1-1
12 1/2	1-0	13 3/8	1-3	12	1-0
14 1/8	1-4	14 1/2	1-12	15 5/8	2-1
14 1/2	1-13	15 1/2	2-3	12 1/2	1-3
12 3/4	1-0	13	1-1	18 1/8	3-5
18	2-15	12 1/2	1-0	16	2-5
14 3/4	1-12	15 1/4	1-11	12 1/8	0-13
14	1-6	12 1/2	1-2	19	3-9
13 1/2	1-3	16 3/4	2-11	12 1/2	0-14
12 1/2	1-2	16 5/8	2-7	13 1/2	1-3
12 3/4	1-0	14 1/4	1-11	12	0-13
12 3/4	1-1	12	0-15	12 1/4	1-0
12 1/8	0-15	14	1-6	14	1-6
13	1-2	12	0-14	12 1/2	1-1
14 1/2	1-14	12 1/2	0-15	12 1/8	0-14
12	1-0	12 1/2	1-1	16	2-1
12	0-14	17 1/4	2-13	14	1-7
12 3/4	1-0	18 3/8	3-8	13 5/8	1-5
14 1/2	1-8	13	1-2	19	4-4
12	0-13	14 1/4	1-9	12	0-13
12 1/4	1-0	13 3/8	1-2	18	3-8
13 1/8	1-2	13	1-3	12 1/4	1-1
18	3-5	12 1/4	0-15	15	1-12
12	1-0	17 1/2	3-1	12 1/2	1-0
16 1/2	2-15	13 1/2	1-1	17 1/8	3-0
12 1/4	0-13	14 3/4	1-12	14 1/4	1-9
13	1-0	20	4-14	12 1/4	0-14
14	1-8	12 1/4	1-0	13 1/2	1-4
19 1/4	3-14	14	1-7	13 1/2	1-5
14 1/4	1-7	14 5/8	1-10	15 1/2	2-0
12 1/8	0-15	14 7/8	1-12		

4
MODEL FITTING

```
REGRESSION COMMANDS:
        REGRESS
        BRIEF
        NOBRIEF
        CONSTANT
        NOCONSTANT

REGRESSION SUBCOMMANDS:
        RESIDS

COLUMN OPERATIONS:
        ABSOLUTE
        SUM
        AVERAGE
        STANDARD DEVIATION
        MEDIAN
        MAXIMUM
        MINIMUM
        COUNT
        DESCRIBE
```

In Chapter 3 we used Minitab to make various transformations on a data set and to plot the resulting transformed data to determine graphically the adequacy of a proposed model. In particular, we learned how to enter data, transform data, obtain a scatterplot, test a proportionality relationship, and estimate the parameters of a model. In this chapter you will learn how to determine the parameters of a model analytically, according to some criterion of "best fit."

For example, suppose it is decided that a parabolic model best explains a behavior being studied, and you are interested in selecting that member of the family $y = Ax^2$ which best fits a given set of data. Using Minitab, you could plot y versus x^2 and graphically estimate the slope of the (approximately) straight line that results if the hypothesized parabolic model is adequate. As another approach you could determine A analytically by selecting an appropriate criterion, such as least-squares, and solving the resulting optimization problem (see Chapter 4, Giordano and Weir, op. cit., for a discussion of model fitting). In this chapter we show how Minitab can be used to solve the least-squares optimization problem, and we analyze the "goodness of fit" of the resulting model.

4.1   Least-squares Curve Fitting: The REGRESS Command

In Minitab the REGRESS command fits a model to a set of data points using the least-squares criterion. That is, given the model $y = f(x)$ and a set of data points, the parameters of the model are determined in such a way that the sum of the squared deviations between the observations and the predictions is minimized. Mathematically, if we are given the set of data points $\{(x_i, y_i): i = 1, 2, \ldots, m\}$, then we wish to <u>minimize</u>

$$S = \sum_{i=1}^{m} [y_i - f(x_i)]^2$$

For example, suppose we wish to fit the model $y = Ax^2$ to the set of data:

x	0.5	1.0	1.5	2.0	2.5
y	0.7	3.4	7.2	12.4	20.1

The least-squares criterion requires the minimization of

$$S = \sum_{i=1}^{5} [y_i - ax_i^2]^2$$

where a is the least-squares estimate of A. Application of the necessary condition yields

$$dS/da = -2 \sum x_i^2 (y_i - ax_i^2) = 0$$

Solving for a we have

$$a = \sum x_i^2 y_i / \sum x_i^4$$

For the data set listed, $\sum x_i^4 = 61.1875$ and $\sum x_i^2 y_i = 195.0$, giving $a = 3.1869$ and the model

$$y = 3.1869x^2$$

The application of the least-squares criterion yields a system of equations, called <u>the normal equations</u>, which must be solved in order to determine the model parameters. For instance, in order to determine the parameters of the model $y = a + bx + cx^2$ which minimize the sum of the squared deviations from the set of data points $\{(x_i, y_i): i = 1, 2, \ldots, m\}$, the following system of equations must be solved:

$$ma + (\Sigma x_i)b + (\Sigma x_i^2)c = \Sigma y_i$$

$$(\Sigma x_i)a + (\Sigma x_i^2)b + (\Sigma x_i^3)c = \Sigma x_i y_i$$

$$(\Sigma x_i^2)a + (\Sigma x_i^3)b + (\Sigma x_i^4)c = \Sigma x_i^2 y_i$$

In Chapter 5 we discuss the solution of a system of linear equations using Minitab. The least-squares criterion is applied more conveniently using the Minitab command REGRESS.

REGRESS

The format for the REGRESS command is

MTB>REGRess **Cj** on **K** predictors in columns **Cj,...,Cj**

The REGRESS command fits the model type

$$Y = b_0 + b_1X_1 + b_2X_2 + \ldots + b_kX_k$$

to the data given in the specified columns. For example, to fit the quadratic model $y = A_0 + A_1x + A_2x^2$ to a data set we could set the observed y values in Column C1, the x values in Column C2, the x^2 values in Column C3, and then use the following command:

REGRess **C1 2 C2 C3**

The REGRESS command is used in conjunction with another Minitab command which controls the amount of statistical information that is provided as output. The statistics requested provide information on the adequacy of the model fit. The following commands control the amount of output provided.

BRIEF (output at level **K**)

The BRIEF command with a specified output level allows the user to select varying degrees of output with the REGRESS command. The level **K** is an integer from 1 to 3, and this part of the command is optional. The larger the value of K, the more output is provided with REGRESS. Specifically:

BRIEF 1: Outputs the regression line, table of coefficients, the mean error S, the coefficient of determination R^2, an adjusted R^2, and part of the analysis of variance table (regression sum of squares, error sum of squares, and the total sum of squares).

BRIEF 2: Outputs, in addition, a full analysis of variance table, and furnishes statistical information on the "unusual" observations.

BRIEF 3: Gives the most complete output. In addition to the previous outputs, it provides statistical information for all the observations.

The default value of the BRIEF command is 2; i.e., BRIEF 2.

NOBRIEF

This command causes "full" output from REGRESS commands which follow. It cancels out the BRIEF command. This command remains in effect until another BRIEF or NOBRIEF command is given.

Exhibit 4.1.1 illustrates fitting a quadratic curve to a given data set and demonstrates the various levels of output described above.

———————————————— EXHIBIT 4.1.1 ————————————————

Fitting A Quadratic With REGRESS

```
MTB > SET Y IN C1
DATA> .7 3.4 7.2 12.4 20.1
DATA> END
MTB > SET X IN C2
DATA> .5:2.5/.5
DATA> END
MTB > LET C3=C2**2
MTB > NAME C1 'Y',C2 'X', C3 'X**2'
MTB > PRINT C1-C3
 ROW       Y       X     X**2

    1     0.7     0.5     0.25
    2     3.4     1.0     1.00
    3     7.2     1.5     2.25
    4    12.4     2.0     4.00
    5    20.1     2.5     6.25

MTB >


MTB > BRIEF
MTB > REGRESS C1 2 C2 C3

THE REGRESSION EQUATION IS
Y = 0.120 - 0.21 X + 3.26 X**2
```

COLUMN	COEFFICIENT	ST. DEV. OF COEF.	T-RATIO = COEF/S.D.
	0.1200	0.6869	0.17
X	-0.211	1.047	-0.20
X**2	3.2571	0.3424	9.51

```
S = 0.3203

R-SQUARED = 99.9 PERCENT
R-SQUARED = 99.8 PERCENT, ADJUSTED FOR D.F.
```

```
ANALYSIS OF VARIANCE

  DUE TO       DF           SS        MS=SS/DF
REGRESSION      2        237.77        118.88
RESIDUAL        2          0.21          0.10
TOTAL           4        237.97


MTB >

MTB > BRIEF 1
MTB > REGRESS C1 2 C2 C3

                              ST. DEV.    T-RATIO =
COLUMN        COEFFICIENT     OF COEF.    COEF/S.D.
                0.1200         0.6869        0.17
X              -0.211          1.047        -0.20
X**2            3.2571         0.3424        9.51

S = 0.3203

R-SQUARED = 99.9 PERCENT

MTB >

MTB > BRIEF 3
MTB > REGRESS C1 2 C2 C3

THE REGRESSION EQUATION IS
Y = 0.120 - 0.21 X + 3.26 X**2

                              ST. DEV.    T-RATIO =
COLUMN        COEFFICIENT     OF COEF.    COEF/S.D.
                0.1200         0.6869        0.17
X              -0.211          1.047        -0.20
X**2            3.2571         0.3424        9.51

S = 0.3203

R-SQUARED = 99.9 PERCENT
R-SQUARED = 99.8 PERCENT, ADJUSTED FOR D.F.

ANALYSIS OF VARIANCE

  DUE TO       DF           SS        MS=SS/DF
REGRESSION      2        237.77        118.88
RESIDUAL        2          0.21          0.10
TOTAL           4        237.97

FURTHER ANALYSIS OF VARIANCE
SS EXPLAINED BY EACH VARIABLE WHEN ENTERED IN THE ORDER GIVEN
  DUE TO       DF           SS
REGRESSION      2        237.77
X               1        228.48
X**2            1          9.28
```

```
MTB > NOBRIEF
MTB > REGRESS C1 2 C2 C3

THE REGRESSION EQUATION IS
Y = 0.120 - 0.21 X + 3.26 X**2

                                    ST. DEV.    T-RATIO =
COLUMN        COEFFICIENT          OF COEF.    COEF/S.D.
                 0.1200             0.6869        0.17
X               -0.211             1.047        -0.20
X**2             3.2571             0.3424        9.51


S = 0.3203

R-SQUARED = 99.9 PERCENT
R-SQUARED = 99.8 PERCENT, ADJUSTED FOR D.F.

ANALYSIS OF VARIANCE

  DUE TO       DF           SS        MS=SS/DF
REGRESSION     2         237.77        118.88
RESIDUAL       2           0.21          0.10
TOTAL          4         237.97

FURTHER ANALYSIS OF VARIANCE
SS EXPLAINED BY EACH VARIABLE WHEN ENTERED IN THE ORDER GIVEN
  DUE TO       DF           SS
REGRESSION     2         237.77
X              1         228.48
X**2           1           9.28

                      Y      PRED. Y    ST.DEV.
ROW       X           Y      VALUE      PRED. Y    RESIDUAL    ST.RES.
  1      0.50      0.700     0.829       0.301     -0.129      -1.19
  2      1.00      3.400     3.166       0.195      0.234       0.92
  3      1.50      7.200     7.131       0.223      0.069       0.30
  4      2.00     12.400    12.726       0.195     -0.326      -1.28
  5      2.50     20.100    19.949       0.301      0.151       1.40


DURBIN-WATSON STATISTIC = 2.64
```

Whenever the REGRESS command fits an equation to the specified columns, it determines the constant b_0. In many modeling applications it is desired that the equation pass through the origin, as in the case of hypothesizing a simple proportionality model. In these situations we wish to suppress the constant b_0.

NOCONSTANT

This command instructs Minitab to fit equations that contain no constant term. Thus an equation of the form

$$Y = b_1X_1 + b_2X_2 + \ldots + b_kX_k$$

is fit to the data given in the columns specified by the REGRESS command. This command remains in effect with all subsequent REGRESS commands unless it is overridden with a CONSTANT command.

CONSTANT

This command instructs Minitab to fit equations that do contain the constant b_0 term. Note that this procedure is the default mode of the REGRESS command. The CONSTANT command remains in effect until overridden with a NOCONSTANT command.

In Exhibit 4.1.2 we illustrate the use of the NOCONSTANT command to fit the model $y = Ax^2$ to the data set given at the beginning of Chapter 4. As obtained previously in Equation (1), the computed model is $y = 3.1869x^2$.

──────────────── EXHIBIT 4.1.2 ────────────────

Fitting A Power Curve With REGRESS

```
MTB > NOBRIEF
MTB > NOCONSTANT
MTB > REGRESS C1 1 C3

THE REGRESSION EQUATION IS
Y = 3.19 X**2
```

COLUMN NOCONSTANT	COEFFICIENT	ST. DEV. OF COEF.	T-RATIO = COEF/S.D.
X**2	3.18693	0.02926	108.92

```
S = 0.2289
```

ANALYSIS OF VARIANCE

DUE TO	DF	SS	MS=SS/DF
REGRESSION	1	621.45	621.45
RESIDUAL	4	0.21	0.05
TOTAL	5	621.66	

ROW	X**2	Y	PRED. Y VALUE	ST.DEV. PRED. Y	RESIDUAL	ST.RES.
1	0.25	0.700	0.797	0.007	-0.097	-0.42
2	1.00	3.400	3.187	0.029	0.213	0.94
3	2.25	7.200	7.171	0.066	0.029	0.13
4	4.00	12.400	12.748	0.117	-0.348	-1.77
5	6.25	20.100	19.918	0.183	0.182	1.32 X

X DENOTES AN OBS. WHOSE X VALUE GIVES IT LARGE INFLUENCE.

DURBIN-WATSON STATISTIC = 2.64

An Illustrative Example: Determining A Constant Of Proportionality

In Exhibit 3.4.1 we tested several proportionality models for predicting the weight of a bass as a function of some easily measurable dimensions. We then visually estimated the constants of proportionality from the graphs. For example, the model $W = 0.0083\ell^3$ was determined graphically. In Exhibit 4.1.3 we illustrate the use of the NOCONSTANT, BRIEF, and REGRESS commands to fit analytically the model $W = k\ell^3$ to the same data set. As illustrated, the least squares estimate of the proportionality constant in the above model is $k = 0.0084368$.

───────────────── EXHIBIT 4.1.3 ─────────────────

Fitting A Cubic Power Curve To The Bass Fishing Derby Data

```
MTB > RETR 'BASS'
MTB > INFO

COLUMN     NAME     COUNT
C1         LENGTH       8
C2         WEIGHT       8

CONSTANTS USED: NONE

MTB > LET C3=C1**3
MTB > NOCONSTANT
MTB > NOBRIEF
MTB > REGRESS C2 1 C3

THE REGRESSION EQUATION IS
WEIGHT = 0.00844 C3
```

COLUMN	COEFFICIENT	ST. DEV. OF COEF.	T-RATIO = COEF/S.D.
NOCONSTANT			
C3	0.0084368	0.0001345	62.72

S = 1.318

73

ANALYSIS OF VARIANCE

DUE TO	DF	SS	MS=SS/DF
REGRESSION	1	6837.8	6837.8
RESIDUAL	7	12.2	1.7
TOTAL	8	6850.0	

ROW	C3	Y WEIGHT	PRED. Y VALUE	ST.DEV. PRED. Y	RESIDUAL	ST.RES.
1	3049	27.000	25.720	0.410	1.280	1.02
2	1953	17.000	16.478	0.263	0.522	0.40
3	5133	41.000	43.305	0.690	-2.305	-2.05R
4	3049	26.000	25.720	0.410	0.280	0.22
5	2012	17.000	16.977	0.271	0.023	0.02
6	5592	49.000	47.181	0.752	1.819	1.68
7	2818	23.000	23.776	0.379	-0.776	-0.61
8	2012	16.000	16.977	0.271	-0.977	-0.76

R DENOTES AN OBS. WITH A LARGE ST. RES.

DURBIN-WATSON STATISTIC = 2.08

4.1 EXERCISES

1. When checking the validity of a proportionality argument, would we use the NOCONSTANT or CONSTANT command? Why?

2. For each of the data sets below, fit the proposed model using the REGRESS command. Test the adequacy of the model by plotting the model predictions and the observations on a single graph. Save your worksheet for future use.

   a. For the bass fishing derby data (see Section 3.4.1, page 47), fit and test the following models: $W \propto g^3$, $W \propto g\ell^2$, $W \propto g^2\ell$.

   b. For the vehicular braking distance model (see Section 3.4.2, page 52 for the model and the data), fit and test the submodel $d_r \propto v$.

   c. For the same model as (b) above, fit and test the submodel $d_b \propto v^2$.

   d. Fit and test a model $e \propto S$ for the following data set:

S	5	10	20	30	40	50	60	70	80	90	100
e	0	19	57	94	134	173	216	256	298	343	380

3. A modeler hypothesizes that the model $P = ae^{bt}$ explains a certain population growth. Employ a logarithmic transformation, and then find ln a and b using the REGRESS command together with the data set below. Find a and use the resultant model to predict P. Use MPLOT to plot the observed and predicted values on a single graph.

t	7	14	21	28	35	42
P	8	41	123	250	280	297

4. The following data represent (hypothetical) energy consumption, normalized to the year 0 (1900). Apply an appropriate transformation to fit and test the model $Q = ae^{bx}$ (see Problem 3 above):

x	0	10	20	30	40	50	60	70
Q	1	2.01	4.06	8.17	16.44	33.12	66.69	134.29

x	80	90	100
Q	270.43	544.57	1096.63

5. In 1601 the German astronomer Johannes Kepler became director of the Prague Observatory. Kepler had been helping Tycho Brahe in collecting 13 years of observations on the relative motion of the planet Mars. By 1609 Kepler had formulated his first two laws:

   i. Each planet moves on an ellipse with the sun at one focus.

   ii. For each planet, the line from the sun to the planet sweeps out equal areas in equal times.

Kepler spent many years verifying these laws and formulating a third law which relates the orbital periods and mean distances from the sun.

   a. Use Minitab to plot the period time T versus the mean distance r using the updated observational data as follows:

Planet	Period (days)	Mean distance from the sun (km x 10^{-6})
Mercury	88	57.9
Venus	225	108.2
Earth	365	149.6
Mars	687	227.9
Jupiter	4329	778.3
Saturn	10753	1427
Uranus	30660	2870
Neptune	60150	4497
Pluto	90670	5907

b.   Assuming a relationship of the form

$$T = Cr^a$$

determine the parameters C and a by plotting ln T versus ln r. Does the model seem reasonable? Can you formulate Kepler's Third law?

4.2 Plotting The Residuals For A Least-Squares Fit

In the previous section you learned how to obtain a least-squares fit of a model, and you plotted the model's predictions on the same graph as the observed data points in order to get a visual indication of how well the model captures the trend of the data. A powerful technique for quickly determining where the model breaks down is to plot the actual deviations or <u>residuals</u> between the observed and predicted values as a function of the independent variable.

1.   The deviations should be randomly distributed and contained in a reasonably small band that is commensurate with the accuracy required by the model.

2.   Any excessively large residual warrants further investigation of the data point in question to discover the cause of the large deviation.

3.   A pattern or trend in the residuals indicates that a predictable effect remains to be modeled, and the nature of the pattern gives clues on how to refine the model, if a refinement is called for.

In this section you will learn how to use the subcommand RESIDS to compute and store the residuals.

Minitab Subcommands; RESIDS

For several commands Minitab offers subcommands. To use a subcommand you must do the following:

1.   Terminate the command line with a semicolon, ";"

2.   Terminate the final subcommand with a period, "."

In particular, REGRESS has many available subcommands. Here we are interested in the subcommand RESIDS which has the following format:

MTB>REGRess Cj on K predictors in columns Cj,...,Cj;

SUBC>RESIDS put into Cj.

Observe that when we terminated the command REGRESS with a semicolon ";" Minitab responded with a SUBC> prompt. The RESIDS subcommand is then terminated with a period.

After fitting a specified model using the least-squares criterion, Minitab stores the differences between the observed and predicted values in the column designated in the RESIDS subcommand. This column can then be plotted, as desired.

You may enter ABORT after any subcommand prompt SUBC>. The ABORT cancels the entire REGRESS command and returns control to the MTB> prompt (in case an error is discovered while entering the RESIDS subcommand).

If you are approximating the least-squares estimators of a model using a transformation, such as $\ln y = \ln a + b \ln x$ to approximate $y = ax^b$, RESIDS computes the deviations <u>from the transformed data</u>, namely $\ln y$ and $\ln x$. To obtain the <u>actual</u> deviations we recommend the following:

1. Determine the actual model $y = ax^b$ from the transformed model.

2. Compute and store the predicted values in a column using the LET command operating on the column containing the observed x values.

3. Compute the residuals as the differences between the observed y values and the predicted y values that were computed and stored in Step 2.

In Exhibit 4.2.1 we demonstrate the use of the RESIDS subcommand to compute, display and plot the residuals for the bass fishing derby model $W \propto \ell^3$ fit in Exhibit 4.1.1. In this case, note that the residuals do not present an obvious trend and the magnitudes of the residuals are fairly reasonable considering the nature of the intended approximation.

———————————————— EXHIBIT 4.2.1 ————————————————

Revisiting The Bass Fishing Derby Cubic Model

```
MTB > RETR 'BASS'
MTB > INFO

COLUMN     NAME       COUNT
C1         LENGTH        8
C2         WEIGHT        8

CONSTANTS USED: NONE

MTB > LET C3=C1**3
MTB > NAME C3 'LNGTH**3'
MTB > NOCONSTANT
MTB > NOBRIEF
MTB > REGRESS C2 1 C3;
SUBC> RESIDS IN C30.
```

THE REGRESSION EQUATION IS
WEIGHT = 0.00844 LNGTH**3

COLUMN	COEFFICIENT	ST. DEV. OF COEF.	T-RATIO = COEF/S.D.
NOCONSTANT			
LNGTH**3	0.0084368	0.0001345	62.72

S = 1.318

ANALYSIS OF VARIANCE

DUE TO	DF	SS	MS=SS/DF
REGRESSION	1	6837.8	6837.8
RESIDUAL	7	12.2	1.7
TOTAL	8	6850.0	

ROW	LNGTH**3	Y WEIGHT	PRED. Y VALUE	ST.DEV. PRED. Y	RESIDUAL	ST.RES.
1	3049	27.000	25.720	0.410	1.280	1.02
2	1953	17.000	16.478	0.263	0.522	0.40
3	5133	41.000	43.305	0.690	-2.305	-2.05R
4	3049	26.000	25.720	0.410	0.280	0.22
5	2012	17.000	16.977	0.271	0.023	0.02
6	5592	49.000	47.181	0.752	1.819	1.68
7	2818	23.000	23.776	0.379	-0.776	-0.61
8	2012	16.000	16.977	0.271	-0.977	-0.76

R DENOTES AN OBS. WITH A LARGE ST. RES.

DURBIN-WATSON STATISTIC = 2.08

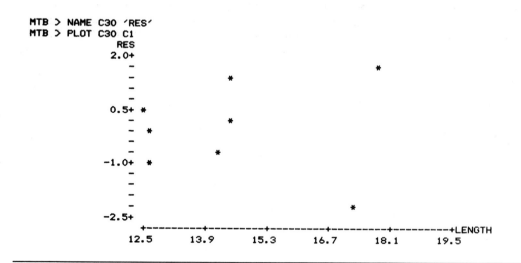

4.2   EXERCISES

1. For each problem in the 4.1 Exercises, complete the following:

    a. Compute and store the residuals.

    b. Plot the residuals versus the independent variable.

    c. Analyze the plot to discern any trends in the residuals.

Save each worksheet for future use.

2. In the data set below, t represents time in hours and P represents the number of yeast cells. Fit a model $P = kt$ to the data, compute and plot the residuals, and discuss how you might adjust your model.

t	0	1	2	3	4	5	6	7	8	9
P	9.6	18.3	29.0	47.2	71.1	119.1	174.6	257.4	350.7	441.0

t	10	11	12	13	14	15	16	17	18
P	513.3	559.7	594.8	629.4	640.8	651.1	655.9	659.6	661.8

4.3   Column Operations Useful For Model Fitting

In using Minitab to fit a model via the least-squares criterion, we typically have considerable information stored in various columns: the observed x and y values, transformed values (if a transformation is needed), y values predicted by the model, residuals, and so forth. In analyzing how well the model fits the data, we may wish to compute various indicators, such as the largest absolute deviation, the sum of the absolute deviations, the average absolute deviation, and so forth. This task can be accomplished easily using the Minitab column operations. We next describe the formats for these column operations and then illustrate their use.

The MEDIAN command computes the median value of the first named column. If the number of values in the column is odd, the median is the value in the middle. If the number of values is even, the median is the average of the 2 middle values. You do not have to order the values of the column in order to use MEDIAN.

The STDE command calculates the sample standard deviation using the formula

$$[1/(n-1) \sum_{i=1}^{N} (x_i - \mu)^2]^{1/2}$$

where N is the number of values in the column and μ is the average value of the numbers in the column.

When using the NCOUNT command, remember a missing value is one that has a missing value code.

Finally, the powerful DESCRIBE command prints descriptive statistics for each of the columns listed. Among the statistics are the NCOUNT, NMISS, MEAN, MEDIAN, STDEV, MAX, and MIN described below. In addition to the MEDIAN, the first quartile Q1 and the third quartile Q3 also are printed. The quartiles Q1, Q2 (the median), and Q3 split the data into four essentially equal parts.

SUMMARY OF COLUMN OPERATIONS

```
ABSOlute value of Ej (put absolute value into Cj)
SUM the values of Cj (put sum into Ej)
AVERage of the values in Cj (put average into Ej)
MEDIan of the values in Cj (put median into Ej)
STDEv of Cj (put standard deviation into Ej)
MAXImum of the values in Cj (put maximum into Ej)
MINImum of the values in Cj (put minimum into Ej)
COUNt the number of values in Cj (put count into Ej)
NCOUnt the number of nonmissing values in Cj (put ncount
    into Ej)
NMISs counts the number of missing values in Cj (put
    nmiss into Ej)
DESCribe columns Cj,...,Cj
```

An Illustrative Example: The Chebyshev Criterion

As discussed above, the REGRESS command determines the parameters of a model such that the sum of the squared residuals is minimized. Other curve-fitting criteria exist although these lead to different optimization problems. For example, the Chebyshev criterion determines the parameters of a model such that the largest absolute deviation is minimized. That is, the Chebyshev criterion determines the parameters of the function type $y = f(x)$ that minimizes the number

$$\text{Maximum } |y_i - f(x_i)|, \quad i = 1, 2, \ldots, m$$

The formulation of a specific problem yields a mathematical program which is either linear or nonlinear. Denote the largest of the residuals that result from a Chebyshev fit by c_{max}. Thus, c_{max} is as small as possible if the Chebyshev criterion is used to determine the model parameters. A bound on c_{max} may be obtained using the results of a least-squares fit as follows:

$$D \leq c_{max} \leq d_{max}$$

where $D = [(d_1{}^2 + d_2{}^2 + \ldots + d_m{}^2)/m]^{1/2}$ and d_{max} is the largest of the residuals d_i resulting when the least-squares criterion is used. Note that both d_{max} and D can readily be obtained in

Minitab using REGRESS, RESIDS, and a few column operations. If the difference between D and d_{max} is large, and minimizing the largest absolute deviation is important in a particular application, one may wish to investigate the Chebyshev criterion further. (See Giordano and Weir, op. cit., for a more thorough discussion of these ideas.)

In Exhibit 4.3.1 we compute the bounds for the bass fishing derby model $W = 0.0084368\ell^3$, which was fit using the REGRESS command in Section 4.1, and find that

$$1.23329 \leq c_{max} \leq 1.8185$$

─────────────────── EXHIBIT 4.3.1 ───────────────────

Computing Bounds For The Deviations In The Bass Fishing Derby Cubic Model

```
MTB > RETR 'BASS'
MTB > LET C3=C2-.0084368*C1**3
MTB > LET C4=C3**2
MTB > LET K1=COUNT (C3)
MTB > LET K2=SUM (C4)
MTB > LET K3=(K2/K1)**.5
MTB > PRINT K3
K3          1.23329
MTB > MAX C3
   MAXIMUM =        1.8185
MTB > NOTE THE BOUNDS ARE 1.23329 AND 1.8185    .
MTB >
```

4.3 EXERCISES

For each of the problems in the 4.2 exercises, compute

  a. d_{max},

  b. D, and

  c. state the bounds for c_{max}.

4.4 Illustrative Examples

Example 1 Vehicular Stopping Distance

In Exhibit 4.4.1 we use the commands presented in this chapter to fit the model for the vehicular braking distance problem originally presented in Section 3.4.2. The model presented consisted of two submodels: one for the reaction distance d_r and the other for the braking distance d_b. The submodels $d_r \propto v$ and $d_b \propto v^2$ are fit to the given data sets for reaction distance and braking distance, respectively. Note that the residual plot shows a trend which indicates that one or both of the submodels are breaking down at high speeds. Further investigation reveals that the submodel for braking distance is inaccurate at high speeds. The bounds for c_{max} are computed as:

$$14.128 \leq c_{max} \leq 28.74 \ .$$

––––––––––––––––––––––––– EXHIBIT 4.4.1 –––––––––––––––––––––––––

Fitting The Vehicular Stopping Distance Submodels

```
MTB > RETR 'BRAKE'
MTB > INFO

COLUMN      NAME        COUNT
C1          SPEED          13
C2          R DIST         13
C3          B DIST         13
C4          T DIST         13
C5          SPEED**2       13
C6          G SPEED        61
C7          P DIST         61

CONSTANTS USED: K1

MTB >
```

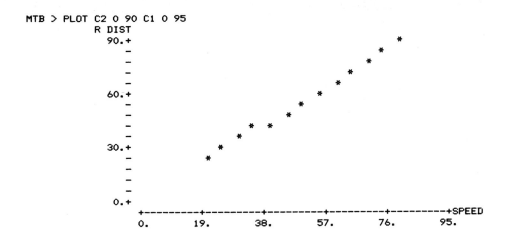

```
MTB > PLOT C2 0 90 C1 0 95
          R DIST
        90.+                                                    *
          -                                              *    *
          -                                           *
          -                                        *
          -                                      *
        60.+                                 *
          -                               *
          -                            *
          -                       *
          -             *    *
        30.+         *
          -       *
          -    *
          -
          -
         0.+
           +---------+---------+---------+---------+---------+SPEED
          0.       19.       38.       57.       76.       95.
```

MTB > NOBRIEF
MTB > NOCON
MTB > REGRESS C2 1 C1;
SUBC> RESIDS IN C30.

THE REGRESSION EQUATION IS
R DIST = 1.10 SPEED

COLUMN	COEFFICIENT	ST. DEV. OF COEF.	T-RATIO = COEF/S.D.
NOCONSTANT			
SPEED	1.10405	0.00142	779.14

S = 0.2728

ANALYSIS OF VARIANCE

DUE TO	DF	SS	MS=SS/DF
REGRESSION	1	45161	45161
RESIDUAL	12	1	0
TOTAL	13	45162	

| | | | Y | PRED. Y | ST.DEV. | | |
ROW	SPEED	R DIST	VALUE	PRED. Y	RESIDUAL	ST.RES.
1	20.0	22.0000	22.0810	0.0283	-0.0810	-0.30
2	25.0	28.0000	27.6012	0.0354	0.3988	1.47
3	30.0	33.0000	33.1215	0.0425	-0.1215	-0.45
4	35.0	39.0000	38.6417	0.0496	0.3583	1.34
5	40.0	44.0000	44.1619	0.0567	-0.1619	-0.61
6	45.0	50.0000	49.6822	0.0638	0.3178	1.20
7	50.0	55.0000	55.2024	0.0709	-0.2024	-0.77
8	55.0	61.0000	60.7227	0.0779	0.2773	1.06
9	60.0	66.0000	66.2429	0.0850	-0.2429	-0.94
10	65.0	72.0000	71.7632	0.0921	0.2368	0.92
11	70.0	77.0000	77.2834	0.0992	-0.2834	-1.12
12	75.0	83.0000	82.8036	0.1063	0.1964	0.78
13	80.0	88.0000	88.3239	0.1134	-0.3239	-1.31

DURBIN-WATSON STATISTIC = 3.37

```
MTB > NAME C30 'RES'
MTB > PLOT C30 VS C1
```

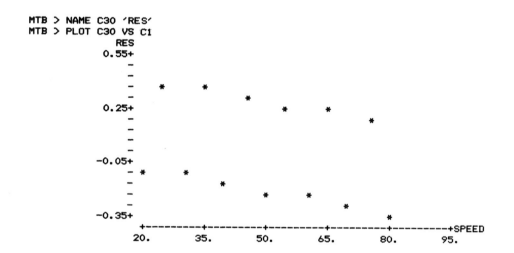

```
MTB > PLOT C3 C5
         B DIST
      450.+
          -                                              *
          -
          -
          -                                         *
      300.+
          -                                    *
          -
          -                               *
          -                          *
      150.+                     *
          -                 *
          -            *
          -       *  *
          -  ** *
       0.+
          +---------+---------+---------+---------+---------+SPEED**2
         0.      1500.    3000.    4500.    6000.    7500.
```

MTB > REGRESS C3 1 C5;
SUBC> RESIDS IN C31.

THE REGRESSION EQUATION IS
B DIST = 0.0542 SPEED**2

COLUMN	COEFFICIENT	ST. DEV. OF COEF.	T-RATIO = COEF/S.D.
NOCONSTANT			
SPEED**2	0.054209	0.001198	45.25

S = 14.79

ANALYSIS OF VARIANCE

DUE TO	DF	SS	MS=SS/DF
REGRESSION	1	447674	447674
RESIDUAL	12	2623	219
TOTAL	13	450297	

		Y	PRED. Y	ST.DEV.		
ROW	SPEED**2	B DIST	VALUE	PRED. Y	RESIDUAL	ST.RES.
1	400	20.00	21.68	0.48	-1.68	-0.11
2	625	28.00	33.88	0.75	-5.88	-0.40
3	900	40.50	48.79	1.08	-8.29	-0.56
4	1225	52.50	66.41	1.47	-13.91	-0.95
5	1600	72.00	86.73	1.92	-14.73	-1.01
6	2025	92.50	109.77	2.43	-17.27	-1.18
7	2500	118.00	135.52	2.99	-17.52	-1.21
8	3025	148.50	163.98	3.62	-15.48	-1.08
9	3600	182.00	195.15	4.31	-13.15	-0.93
10	4225	220.50	229.03	5.06	-8.53	-0.61
11	4900	266.00	265.62	5.87	0.38	0.03
12	5625	318.00	304.92	6.74	13.08	0.99
13	6400	376.00	346.94	7.67	29.06	2.30RX

R DENOTES AN OBS. WITH A LARGE ST. RES.
X DENOTES AN OBS. WHOSE X VALUE GIVES IT LARGE INFLUENCE.

DURBIN-WATSON STATISTIC = 0.22

MTB > NAME FOR C31 'RESI'
MTB > PLOT C31 VS C1

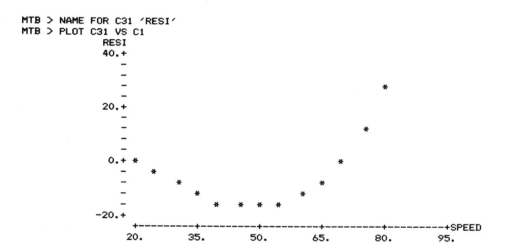

```
MTB > NOTE FROM THE PREVIOUS REGRESSIONS WE FOUND THE SLOPES 1.1 AND .054   .
MTB > LET C8=1.1 *C6 +.054*C6**2
MTB > NAME FOR C8 IS 'PRED DIS'

MTB > WIDTH 85
MTB > MPLOT C4 C1 C8 C6
```

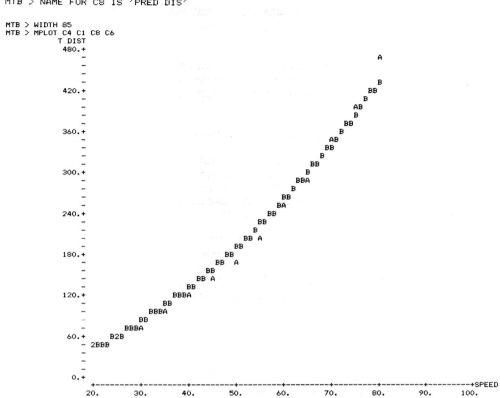

```
MTB > LET C21=1.10405*C1+.054209*C1**2
MTB > LET C22=C4-C21
MTB > NAME  C21 'PRED D', C22 'DEV'
MTB > PRINT C4 C21 C22
  ROW  T DIST    PRED D        DEV

    1    42.0    43.765    -1.7645
    2    56.0    61.482    -5.4818
    3    73.5    81.909    -8.4095
    4    91.5   105.048   -13.5477
    5   116.0   130.896   -14.8962
    6   142.5   159.455   -16.9553
    7   173.0   190.725   -17.7247
    8   209.5   224.704   -15.2045
    9   248.0   261.395   -13.3948
   10   292.5   300.795    -8.2952
   11   343.0   342.906     0.0936
   12   401.0   387.728    13.2723
   13   464.0   435.260    28.7399

MTB > NOTE OBTAIN THE BOUNDS ON THE MODEL.
MTB > LET C23=C22**2
MTB > LET K1=COUNT (C22)
MTB > LET K2= SUM (C23)
MTB > LET K3=(K2/K1)**.5
MTB > PRINT K3
K3        14.1218
MTB > MAX C22
   MAXIMUM =       28.740
MTB > NOTE THE BOUND ARE 14.1218 AND 28.740  .
```

Example 2   The Elapsed Time Of A Tape Recorder

In Exhibit 4.4.2 we fit and analyze the goodness of fit of the model for the elapsed time of a tape recorder discussed in Section 3.4.3. The submodel $c \propto \theta/2\pi$ is tested and the constant of proportionality ascertained. Ultimately, the model

$$t = k_2 c + k_3 c^2$$

is fit using the data provided. The fit of the model is tested and found to be reasonable. The bounds for c_{max} are found to be

$$0.338096 \leq c_{max} \leq 0.50427.$$

EXHIBIT 4.4.2
Fitting The Tape Recorder Model

```
MTB > SET C1
DATA> 100:800/100
DATA> END
MTB > SET C2
DATA> 205 430 677  945 1233 1542 1872 2224
DATA> END
MTB > NAME C1 'COUNTER', C2 'TIME'



MTB > PRINT C1 C2
  ROW   COUNTER    TIME

    1     100       205
    2     200       430
    3     300       677
    4     400       945
    5     500      1233
    6     600      1542
    7     700      1872
    8     800      2224

MTB > LET C3=C1**2
MTB > NAME C3 'C**2'



MTB > REGRESS C2 2 C1 C3;
SUBC> RESIDS IN C30.

THE REGRESSION EQUATION IS
TIME = 1.94 COUNTER + 0.00105 C**2
```

		ST. DEV.	T-RATIO =
COLUMN	COEFFICIENT	OF COEF.	COEF/S.D.
NOCONSTANT			
COUNTER	1.94293	0.00110	1761.19
C**2	0.00104578	0.00000168	621.61

```
S = 0.3904
```

ANALYSIS OF VARIANCE

DUE TO	DF	SS	MS=SS/DF
REGRESSION	2	13926892	6963446
RESIDUAL	6	1	0
TOTAL	8	13926892	

FURTHER ANALYSIS OF VARIANCE
SS EXPLAINED BY EACH VARIABLE WHEN ENTERED IN THE ORDER GIVEN

DUE TO	DF	SS
REGRESSION	2	13926892
COUNTER	1	13867988
C**2	1	58902

ROW	COUNTER	Y TIME	PRED. Y VALUE	ST.DEV. PRED. Y	RESIDUAL	ST.RES.
1	100	205.00	204.75	0.09	0.25	0.66
2	200	430.00	430.42	0.16	-0.42	-1.17
3	300	677.00	677.00	0.19	0.00	0.00
4	400	945.00	944.50	0.19	0.50	1.48
5	500	1233.00	1232.91	0.18	0.09	0.26
6	600	1542.00	1542.24	0.17	-0.24	-0.67
7	700	1872.00	1872.48	0.21	-0.48	-1.45
8	800	2224.00	2223.64	0.31	0.36	1.53

DURBIN-WATSON STATISTIC = 2.09

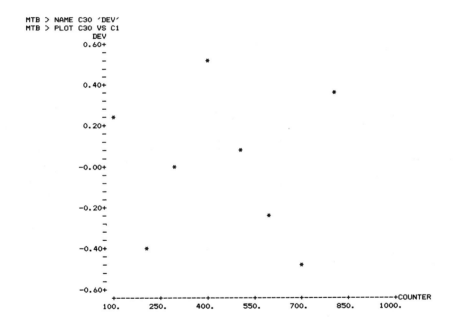

```
MTB > NAME C30 'DEV'
MTB > PLOT C30 VS C1
```

90

```
MTB > SET COUNTERS IN C10
DATA> 100:800/10
DATA> END


MTB > LET C11=1.94293*C10+.00104578*C10**2
MTB > WIDTH 80
MTB > MPLOT C2 C1 C11 C10
        TIME
      2400.+
           -
           -
           -
           -                                                              2
           -                                                            BB
      2100.+                                                          BB
           -                                                        BB
           -                                                      BB
           -                                                     B
           -                                                   BB
           -                                                 B2
      1800.+                                                BB
           -                                               B
           -                                             BB
           -                                           BB
           -                                          2B
      1500.+                                        BB
           -                                      BB
           -                                     BB
           -                                   2B
           -                                 BB
      1200.+                               BB
           -                             BB
           -                           BB
           -                         2B
      900.+                         BB
           -                      BB
           -                    BBB
           -                   BB
           -                 B2
      600.+                BBB
           -             BB
           -           BBB
           -          B2
           -        BBB
      300.+     BBB
           -   BB
           - 2
           -
           -
        0.+
           +---------+---------+---------+---------+---------+---------+---------+---------+---------+COUNTER
          100.      200.      300.      400.      500.      600.      700.      800.      900.
```

MTB > NOTE :FINDING THE BOUNDS.
MTB > LET K1=COUNT (C30)
MTB > LET C31=C30**2
MTB > LET K2=SUM (C31)
MTB > LET K3=(K2/K1)**.5
MTB > PRINT K3
K3   0.338106
MTB > MAX C30
   MAXIMUM =   0.50427
MTB > NOTE THE BOUNDS ARE .338096 AND .50427 .

4.4 PROJECTS

For each of the projects listed at the end of Chapter 3, use Minitab to:

  a. Construct an explicative model addressing your scenario.

  b. Use a scatterplot to help appraise the accuracy of the data provided. How accurate do you think the data are?

  c. Use a graphical plot to obtain an initial test of any sub-models that use a proportionality relationship. Estimate visually the constant of proportionality.

  d. Obtain a least-squares fit of the model using REGRESS and RESIDS.

  e. Use MPLOT to display the model predictions and observations simultaneously. How well does your model fit the data? Let width = 100 and height = 50 for adequate resolution.

  f. Plot the residuals versus the independent variable. Where does the model not fit well? Is there a trend in the residuals?

  g. Find bounds for c_{max}:

$$D \leq c_{max} \leq d_{max}$$

  h. Refine your model if necessary.

5

EMPIRICAL MODEL CONSTRUCTION

```
MATRIX ENTRY:
     READ

MATRIX OPERATIONS:
     INVERT
     MULTIPLY

PROGRAM STORAGE AND EXECUTION:
     STORE
     EXECUTE
```

In previous chapters we assumed the modeler has developed a reasoning to relate the variables under consideration and explain, in some sense, the observed behavior. If collected data then corroborate the reasonableness of the assumptions underlying the hypothesized relationships, the parameters of the model can be chosen that "best fit" the model type to the collected data according to some criterion (such as least-squares or the Chebyshev criterion).

In many practical cases the modeler is unable to construct a tractable model form that satisfactorily explains the behavior. Nevertheless, if it is necessary to predict the behavior, the modeler may conduct experiments (or otherwise gather data) to investigate the behavior of the dependent variable(s) for selected values of the independent variable(s) within some range of interest. In essence the modeler desires to construct an empirical model based on the collected data. In this situation the modeler is strongly influenced by the data that have been carefully collected and analyzed, so he or she seeks a curve that captures the trend of the data in order to predict between the collected data points.

To contrast explicative and empirical models, consider the data shown in Figure 5-1a. If the modeler's assumptions lead to the expectation of a quadratic explicative model, a parabola would be fit to the data points, as illustrated in Figure 5-1b. However, if the modeler has no reason to expect a model of a particular type, a smooth curve may be passed through the data points to serve as an empirical model, as illustrated in Figure 5-1c.

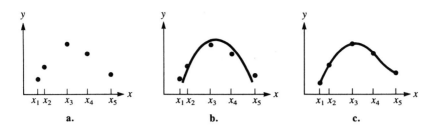

FIGURE 5-1. If the modeler expects a quadratic relationship, a parabola may be fit to the data, as in b. Otherwise a smooth curve may be passed through the points, as in c.

In this chapter we address the construction of empirical models. We begin by investigating simple one-term models which capture the trend of the data. Then we consider higher-order polynomials that pass through the collected data points. Next, the "smoothing" of data with low-order polynomials is investigated. Finally, we present the technique of cubic spline interpolation, which passes a distinct cubic polynomial across each successive pair of data points to form a single smooth curve. While each of these techniques has its own advantages and disadvantages, which may justify its use for a specific application, the flow chart in Figure 5-2 presents a logical procedure for determining an appropriate empirical model. As portrayed in the figure, we begin with a scatterplot of the data to determine if a trend exists. If a trend is discernible, we start with the simplest technique available and gradually increase the sophistication until an empirical model is developed that satisfies the requirements of the particular application. For a more detailed discussion of empirical model construction, see Chapter 6 of Giordano and Weir, op. cit.

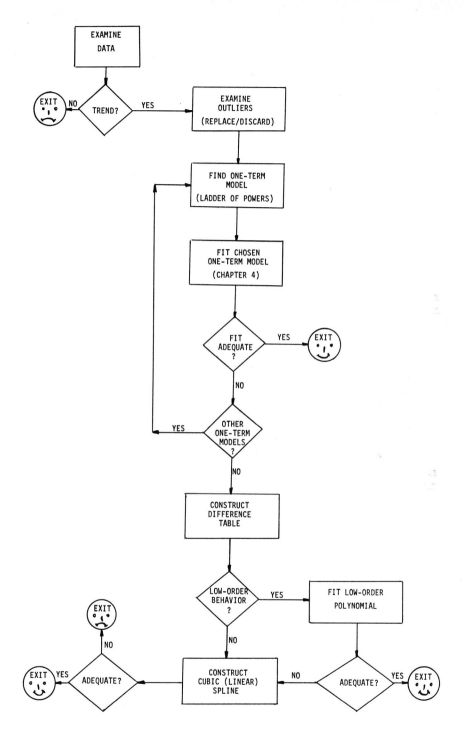

FIGURE 5-2. If a trend in the data is discernible, the simplest model which captures the trend of the data is selected.

5.1 One-Term Models

We commence our study of empirical model building with the presentation of simple one-term models. These models have several advantages. One obvious and powerful advantage is their simplicity. Another is that a one-term model may capture the trend of the data better than any other empirical model (such as a polynomial). This feature is particularly important in situations when the modeler intends to make predictions by extrapolating for values outside the intervals of the observations; for instance, when the modeler intends to predict a future value based on historical information.

When constructing an empirical model, always begin with a careful analysis of the collected data. Investigate whether the data suggest the existence of a trend. Are there data points that fail to follow the trend? If such "outliers" do exist, you may wish to discard them. Of if they were obtained experimentally, you may choose to repeat the experiment as a check for a data collection error. Once it seems evident that a trend does exist, the use of one-term models is investigated using either of the following two techniques:

1. The matching of the trend's concavity using the "Ladder of Transformations."

2. The linearization of the data with an appropriate power transformation.

If a one-term model is chosen, we then estimate the parameters of the model graphically, as in Chapter 3, or analytically, as in Chapter 4. Eventually we must analyze how well the model fits the data. Because of their inherent simplicity, we caution that one-term models cannot possibly fit all data sets, so we will have to study other methods as well.

The Ladder Of Transformations

Suppose we plot our data and observe the existence of a simple concave-up trend like this:

Or perhaps the trend is concave-down like this:

In such cases an appropriate model may be selected from among those listed in Table 5-1. The idea of the table is to match the concavity of the observed trend to an appropriate power. Note that $y = z$ represents no concavity (i.e., a straight line). As we increase the power of z and walk up the ladder, we increase the concavity; as we decrease the power of z and walk down the ladder, we decrease the concavity. We assume here that the trend of the data is always increasing.

Since the powers $1/z^{1/2}$, $1/z$ or $1/z^2$ change an increasing function into a decreasing one, we compensate for this by including minus signs in the table. The variable z in the table may refer either to the independent or the dependent variable of your data.

TABLE 5-1

The Ladder Of Transformations

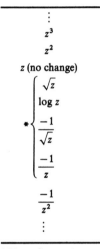

* Most often used trans-
formations

Once a candidate model is chosen from the ladder, the model can be tested qualitatively in several ways. First, we can transform the data as suggested by the model and then plot the transformed data. If the chosen model is reasonable, the graph should approximate a straight line passing through the origin. If we are satisfied with this qualitative test, the constant of proportionality can be estimated graphically using the techniques of Chapter 3, or analytically, using one of the methods in Chapter 4. Once the constant of proportionality is known, we can plot the observations and model predictions on a single graph to determine if we have enough concavity. Alternatively, we can plot the residuals to determine if more or less concavity will improve the fit. These procedures can be implemented conveniently using Minitab, as illustrated by the next example.

Example 1 The Bass Fishing Derby

Consider the Bass Fishing Derby discussed previously in Section 3.4. In Exhibit 5.1.1, the scatterplot of the original data suggests a trend which is concave up. Thus we select the z^2 transformation and plot W vs ℓ^2. Since the plot is reasonably linear, we fit the model analytically (using least-squares) and plot the residuals. Since there is an evident trend in the residuals, we investigate W vs ℓ^3 in a similar manner. The two resulting models are

$$W = 0.131\ell^2$$

and

$$W = 0.00844\,\ell^3.$$

Note that the residuals of W vs ℓ^3 are more random about the origin, indicating that the second model is the better one.

──────────────────── EXHIBIT 5.1.1 ────────────────────

The Bass Fishing Derby Revisited, Using the Ladder of Transformations

```
MTB > RETR 'BASS'
MTB > INFO

COLUMN    NAME       COUNT
C1        LENGTH        8
C2        WEIGHT        8

CONSTANTS USED: NONE


MTB > PLOT C2 C1
          WEIGHT
          60.+
             -
             -
             -
             -                                           *
          45.+
             -                                     *
             -
             -
             -
          30.+
             -                 2
             -              *
             -
             - **
          15.+ *
             +---------+---------+---------+---------+---------+LENGTH
            12.5      13.9      15.3      16.7      18.1      19.5
```

98

```
MTB > PLOT C2 0 60 C1 0 20
```

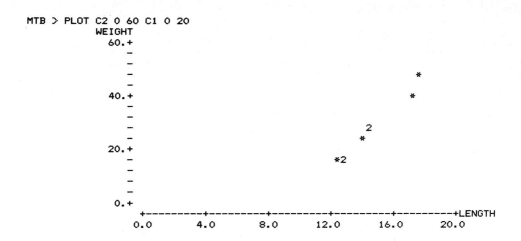

```
MTB > LET C3=C1**2
MTB > NAME C3 'LNGTH**2'
MTB > PLOT C2 0 60 C3 0 400
```

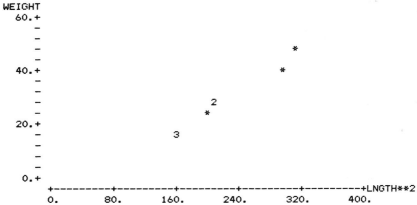

```
MTB > NOCONS
MTB > NOBRIEF
MTB > REGRESS C2 1 C3;
SUBC> RESIDS IN C10.

THE REGRESSION EQUATION IS
WEIGHT = 0.131 LNGTH**2
```

		ST. DEV.	T-RATIO =
COLUMN	COEFFICIENT	OF COEF.	COEF/S.D.
NOCONSTANT			
LNGTH**2	0.131084	0.006815	19.23

```
S = 4.263
```

ANALYSIS OF VARIANCE

DUE TO	DF	SS	MS=SS/DF
REGRESSION	1	6722.8	6722.8
RESIDUAL	7	127.2	18.2
TOTAL	8	6850.0	

ROW	LNGTH**2	Y WEIGHT	PRED. Y VALUE	ST.DEV. PRED. Y	RESIDUAL	ST.RES.
1	210	27.00	27.56	1.43	-0.56	-0.14
2	156	17.00	20.48	1.06	-3.48	-0.84
3	298	41.00	39.01	2.03	1.99	0.53
4	210	26.00	27.56	1.43	-1.56	-0.39
5	159	17.00	20.89	1.09	-3.89	-0.94
6	315	49.00	41.30	2.15	7.70	2.09R
7	200	23.00	26.15	1.36	-3.15	-0.78
8	159	16.00	20.89	1.09	-4.89	-1.19

R DENOTES AN OBS. WITH A LARGE ST. RES.

DURBIN-WATSON STATISTIC = 2.45

MTB > PLOT C10 VS C1

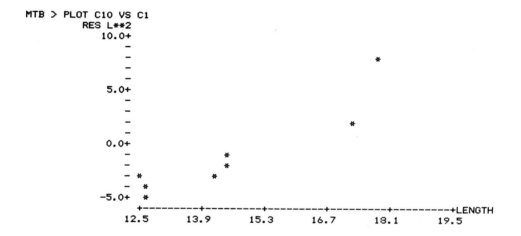

```
MTB > LET C4=C1**3
MTB > NAME C4 'LNGTH**3'
MTB > PLOT C2 0 60 VS C4 0  8000
```

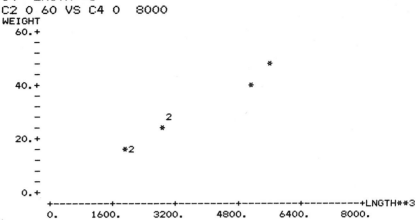

```
MTB > REGRESS C2 1 C4;
SUBC> RESIDS IN C11.

THE REGRESSION EQUATION IS
WEIGHT = 0.00844 LNGTH**3
```

		ST. DEV.	T-RATIO =
COLUMN	COEFFICIENT	OF COEF.	COEF/S.D.
NOCONSTANT			
LNGTH**3	0.0084368	0.0001345	62.72

$S = 1.318$

ANALYSIS OF VARIANCE

DUE TO	DF	SS	MS=SS/DF
REGRESSION	1	6837.8	6837.8
RESIDUAL	7	12.2	1.7
TOTAL	8	6850.0	

ROW	LNGTH**3	WEIGHT	PRED. Y VALUE	ST.DEV. PRED. Y	RESIDUAL	ST.RES.
1	3049	27.000	25.720	0.410	1.280	1.02
2	1953	17.000	16.478	0.263	0.522	0.40
3	5133	41.000	43.305	0.690	-2.305	-2.05R
4	3049	26.000	25.720	0.410	0.280	0.22
5	2012	17.000	16.977	0.271	0.023	0.02
6	5592	49.000	47.181	0.752	1.819	1.68
7	2818	23.000	23.776	0.379	-0.776	-0.61
8	2012	16.000	16.977	0.271	-0.977	-0.76

```
R DENOTES AN OBS. WITH A LARGE ST. RES.

DURBIN-WATSON STATISTIC = 2.08

MTB > NAME C11 'RES L**3'
```

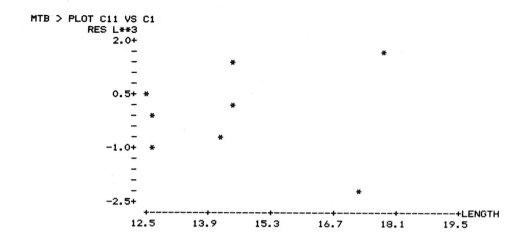

```
MTB > PLOT C11 VS C1
      RES L**3
      2.0+
         -
         -                                          *
         -                     *
         -
      0.5+  *
         -
         -    *                 *
         -
         -                *
     -1.0+    *
         -
         -
         -
         -
     -2.5+                               *
         +---------+---------+---------+---------+---------+LENGTH
        12.5      13.9      15.3      16.7      18.1      19.5
```

Power Transformations

Observe that most of the models in the Ladder of Transformations are "power models" of the form $y = ax^b$. A candidate power
model can be obtained in another manner. As an alternative to
using the Ladder of Transformations, transform the general power
model $y = ax^b$ by directly applying the logarithm to both sides of
the equation. Then use the data points to find the "best" estimates of a and b according to some criterion (such as least-
squares). The estimate of the exponent b can be rounded if
appropriate. Not all of the models in the Ladder of Transformations are obtainable by this process. We also allow noninteger
exponents for the power models.

The above procedure is readily implemented using Minitab.
After a scatterplot reveals a trend, test the model

$$\ln y = \ln a + b \ln x$$

with the data and estimate ln a (and a) and b either graphically or
analytically. Using procedures developed in previous chapters we
can then test the adequacy of the model. We illustrate the procedure for the Bass Fishing Derby in the next exhibit and obtain the
model

$$W = 0.00804\,\ell^{3.02}.$$

102

———— EXHIBIT 5.1.2 ————

The Bass Fishing Derby Revisited,
Using a Power Transformation

```
MTB > LOGE C1 PUT IN C5
MTB > LOGE C2 PUT IN C6
MTB > NAME C5 'LN L', C6 'LN W'
MTB > PRINT C5 C6
 ROW      LN L        LN W

   1     2.67415     3.29584
   2     2.52573     2.83321
   3     2.84781     3.71357
   4     2.67415     3.25810
   5     2.53568     2.83321
   6     2.87638     3.89182
   7     2.64795     3.13549
   8     2.53568     2.77259
```

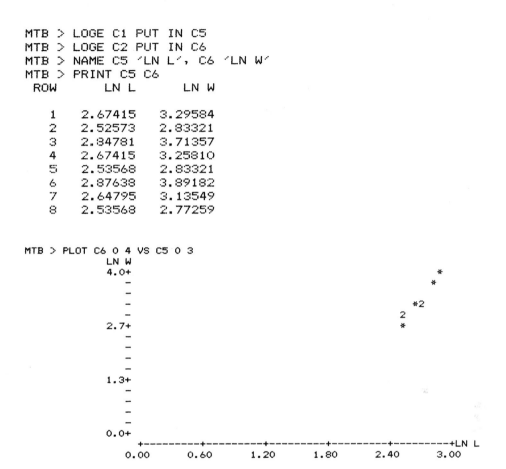

```
MTB > PLOT C6 0 4 VS C5 0 3
          LN W
         4.0+                                              *
            -                                            *
            -
            -                                        *2
            -                                      2
         2.7+                                      *
            -
            -
            -
         1.3+
            -
            -
            -
         0.0+
            +---------+---------+---------+---------+---------+LN L
          0.00      0.60      1.20      1.80      2.40      3.00
```

```
MTB > CONS
MTB > REGRESS C6 1 C5

THE REGRESSION EQUATION IS
LN W = - 4.82 + 3.02 LN L

                                  ST. DEV.      T-RATIO =
COLUMN        COEFFICIENT        OF COEF.      COEF/S.D.
               -4.8230            0.3347        -14.41
LN L            3.0171            0.1255         24.05

S = 0.04548

R-SQUARED = 99.0 PERCENT
R-SQUARED = 98.8 PERCENT, ADJUSTED FOR D.F.

ANALYSIS OF VARIANCE

 DUE TO       DF            SS        MS=SS/DF
REGRESSION    1          1.1962       1.1962
RESIDUAL      6          0.0124       0.0021
TOTAL         7          1.2086

                      Y      PRED. Y    ST.DEV.
ROW     LN L        LN W     VALUE     PRED. Y    RESIDUAL    ST.RES.
 1      2.67       3.2958    3.2453    0.0161      0.0506      1.19
 2      2.53       2.8332    2.7975    0.0237      0.0358      0.92
 3      2.85       3.7136    3.7692    0.0280     -0.0557     -1.55
 4      2.67       3.2581    3.2453    0.0161      0.0128      0.30
 5      2.54       2.8332    2.8275    0.0228      0.0057      0.15
 6      2.88       3.8918    3.8554    0.0310      0.0364      1.09
 7      2.65       3.1355    3.1662    0.0162     -0.0307     -0.72
 8      2.54       2.7726    2.8275    0.0228     -0.0549     -1.40


DURBIN-WATSON STATISTIC = 1.56

MTB >
```

104

```
MTB > LET K1=EXPO(-4.8230)
MTB > PRINT K1
K1        0.00804261
MTB > NOTE THE MODEL IS  W= .00804621 * L ** 3.0171  .
MTB > LET C7 = .00804621 * C1 ** 3.0171
MTB > NAME C7 'PRED W'
MTB > LET C8= C2-C7
MTB > NAME C8 'RES  TL'
MTB > PRINT C2 C7 C8
 ROW   WEIGHT     PRED W    RES   TL

  1       27     25.6775    1.32247
  2       17     16.4088    0.59119
  3       41     43.3617   -2.36169
  4       26     25.6775    0.32247
  5       17     16.9089    0.09113
  6       49     47.2657    1.73431
  7       23     23.7258   -0.72577
  8       16     16.9089   -0.90887


MTB > PLOT C8 VS C1
        RES  TL
        2.0+
          -                                         *
          -                  *
          -
          -
        0.5+ *
          -    *               *
          -
          -
          -              *
       -1.0+   *
          -
          -
          -
       -2.5+                              *
          +---------+---------+---------+---------+---------+LENGTH
         12.5      13.9      15.3      16.7      18.1      19.5
```

5.1 EXERCISES

1. Use the Ladder of Transformations to determine if a simple one-term model captures the trend of the data in the following problems. If a one-term model does seem reasonable, determine the parameters of the model using the least-squares criterion and Minitab.

   a. From 4.1 Exercises, page 74,

      (1) Problem 4.1.2

      (2) Problem 4.1.3

      (3) Problem 4.1.4

      (4) Problem 4.1.5

   b. From 4.2 Exercises, page 79, Problem 4.2.2.

   c. Relate volume V as a function of diameter D for the following set of data:

D	36	28	28	41	19	32	22	38	25	17	31
V	192	113	88	294	28	123	51	252	56	16	141

D	20	25	19	39	33	17	37	23	39
V	32	86	21	231	187	22	205	57	265

   d. Relate x as a function of y for the following data set:

x	35.97	67.21	92.96	141.7	483.7	886.7
y	0.241	0.615	1.0	1.881	11.86	29.46

x	1783.0	2794	3666
y	84.01	164.8	248.4

   e. Relate weight W as a function of length ℓ :

ℓ	12.5	12.625	14.125	14.5	17.25	17.75
W	17	16.5	23	26.5	41	49

   f. Relate the sheep population P in Tasmania as a function of the year t:

t	1814	1824	1834	1844	1854	1864
P	125	275	830	1200	1750	1650

   g. Relate the walking speed V to the population P:

P	365	2500	5491	14000	23700	49375	70700	78200	138000
V	2.76	2.27	3.31	3.7	3.27	4.9	4.31	3.85	4.39

P	304500	341948	867023	1092759	1340000	2602000
V	4.42	4.81	5.21	5.88	5.62	5.05

106

2. Apply the power transformation method to determine a power model for the problems listed in Exercise 1 above. If a power model does appear reasonable, determine the parameters of the model using the least-squares criterion.

3. Plot the residuals for the models listed in Exercises 1 and 2 above.

5.2   Fitting An N-1 Order Polynomial To N Data Points

Because of their inherent simplicity, one-term models facilitate model analysis including sensitivity analysis, optimization, estimation of rates of change and area under the curve applications, and so forth. However, precisely because of this simplicity, one-term models are not likely to capture the trend of an arbitrary data set. In many cases models with more than one term must be considered. The remainder of this chapter considers one type of multiterm model; namely, the polynomial. Since polynomials are easy to integrate and differentiate, they are especially popular to use. We begin by studying polynomials that pass through each point in a data set containing only one observation for each value of the independent variable.

Consider passing a quadratic $P_2(x) = a_0 + a_1x + a_2x^2$ through the following data points:

x_i	1	2	3
y_i	5	8	25

Requiring that $P_2(x_i) = y_i$ yields the following system of equations:

$$a_0 + 1a_1 + 1a_2 = 5$$

$$a_0 + 2a_1 + 2^2a_2 = 8$$

$$a_0 + 3a_1 + 3^2a_2 = 25$$

To solve the above system conveniently using Minitab, we consider the linear system in the form of the matrix equation

$$AX = B$$

which has the solution $X = A^{-1}B$, provided that A is invertible. Thus, the above system can be written in matrix form as

$$\begin{bmatrix} 1 & 1 & 1 \\ 1 & 2 & 4 \\ 1 & 3 & 9 \end{bmatrix} \begin{bmatrix} a_0 \\ a_1 \\ a_2 \end{bmatrix} = \begin{bmatrix} 5 \\ 8 \\ 25 \end{bmatrix}$$

and, since the coefficient matrix is invertible, the solution is

$$\begin{bmatrix} a_0 \\ a_1 \\ a_2 \end{bmatrix} = \begin{bmatrix} 3 & -3 & 1 \\ -2.5 & 4 & -1.5 \\ 0.5 & -1 & 0.5 \end{bmatrix} \begin{bmatrix} 5 \\ 8 \\ 25 \end{bmatrix} = \begin{bmatrix} 16 \\ -18 \\ 7 \end{bmatrix}$$

or $P_2(x) = 16 - 18x + 7x^2$.

In general, the requirement that an n-1 degree polynomial pass through n distinct data points (because only one observation is allowed for each value of the independent variable) yields a system of n linear algebraic equations in n unknowns. It is important to realize that large systems of equations can be difficult to solve with great accuracy, and small round-off errors in computer arithmetic can cause large oscillations to occur due to the presence of the higher-order terms. (By "large" we mean more than 13 equations or so, depending on the coefficient matrix--the size of the range of values of the coefficients, from very small to very large, or the number of zeros scattered throughout the matrix and where they occur. These are the kinds of features that create difficulties.)

Although polynomials are "perfect" fitting models in the sense of any of the curve-fitting criteria studied in Chapter 4 (because they have zero sum of deviations, zero sum of squared deviations, and zero largest absolute deviation), they do have the following unpleasant features which interfere with their ability to capture the trend of the data:

1. "Snaking"

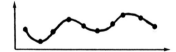

2. "Oscillation"

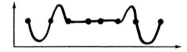

Because of these troublesome behaviors associated with higher-order polynomials, we will study the idea of fitting low-order polynomials in the next section. Low-order polynomials effectively smooth the data (since a low-order polynomial is not required to pass exactly through all the data points).

Using Minitab to determine the coefficients of the n-1 order polynomial passing through the n data points requires first that we obtain the matrix equation $AX = B$. Then we need to be able to do the following:

---

1.   Enter the A and B matrices.

2.   Determine A^{-1}, the inverse of matrix A.

3.   Premultiply B by A^{-1}.

We begin by presenting the Minitab commands READ, INVERT, and MULTIPLY. The format for the READ command is:

READ n by **n** matrix into **Mj**

For example:

```
MTB>READ 2 by 2 matrix into M1
DATA>1   2
DATA>3   4
MTB>PRINt M1
```

Each DATA> command represents a <u>row</u> of the matrix. Thus the matrix is entered one row at a time, and the above sequence of commands results in the following output:

MATRIX M1

        1   2

        3   4

The format for the INVERT command is:

MTB>**INVErt Mj** put into **Mj**

The format for the MULTIPLY command is:

MTB>**MULTiply Mj** by **Mj** put into **Mj**

When using the MULTIPLY command, matrices are multiplied in the same order as given. We illustrate the above commands in Exhibit 5.2.1 by solving the matrix equation

$$\begin{bmatrix} 1 & 1 & 1 \\ 1 & 2 & 4 \\ 1 & 3 & 9 \end{bmatrix} \begin{bmatrix} a_0 \\ a_1 \\ a_2 \end{bmatrix} = \begin{bmatrix} 5 \\ 8 \\ 25 \end{bmatrix}$$

First we enter the A and B matrices:

```
MTB>READ  3  3  M1
DATA>1  1  1
DATA>1  2  4
DATA>1  3  9
MTB>READ  3  1  M2
DATA>5
DATA>8
DATA>25
```

Next we invert the A matrix to obtain A^{-1} and store it as M3:

```
MTB>INVErt M1  M3
```

Now premultiply the matrix B (= M2) by A^{-1} (= M3) and store the result in M4:

```
MTB>MULT  M3  M2  M4
```

Finally, print out the result $X = A^{-1}B$:

```
MTB>PRIN M4
```

The solution, as presented in Exhibit 5.2.1, is

$$y = 16 - 18x + 7x^2.$$

———————————————————— EXHIBIT 5.2.1 ————————————————————

Solving A Square System Of Linear Algebraic Equations

```
MTB > READ 3 3 MATRIX INTO M1
DATA> 1 1 1
DATA> 1 2 4
DATA> 1 3 9
     3 ROWS READ
MTB > READ 3 1  MATRIX M2
DATA> 5
DATA> 8
DATA> 25
     3 ROWS READ
MTB > INVERT M1 M3
MTB > MULT M3 M2 M4
MTB > PRINT M4
MATRIX M4

   16
  -18
    7
```

```
MTB > NOTE THE COEFFICIENTS ARE 16,-18 AND 7.
MTB > SET C1
DATA> 1 2 3
DATA> END
MTB > LET C2=16-18*C1+7*C1**2
MTB > NAME C1 'X',C2 'Y'
MTB > PRINT C1 C2
 ROW    X      Y

   1    1      5
   2    2      8
   3    3     25


MTB > PLOT C2 C1
         Y
     27.0+                                         *
         -
         -
         -
         -
     18.0+
         -
         -
         -
         -
      9.0+
         -                      *
         - *
         -
         -
      0.0+
         +---------+---------+---------+---------+---------+X
        1.00      1.50      2.00      2.50      3.00      3.50
MTB > NOTE THAT THE PREDICTIONS AGREE WITH THE OBSERVATIONS.
```

Example   The Elapsed Time Of A Tape Recorder

Suppose we wish to construct an empirical model to predict the amount of elapsed playing time of a tape recorder as a function of its counter reading (see Giordano and Weir, op. cit., page 74 for an explicative model). Assume the following data have been collected, where c_i is the counter reading and t_i is the elapsed time in seconds:

c_i	100	200	300	400	500	600	700	800
t_i	205	430	677	945	1233	1542	1872	2224

In Exhibit 5.2.2 a polynomial is fit to these data points using Minitab. The resulting polynomial is given by

$$t = P_7(c) = -0.13927 + 2.32723c - 0.28895c^2 + 0.19723c^3 -$$

$$0.05338c^4 + 0.00800c^5 - 0.00062c^6 + 0.00002c^7$$

111

This polynomial can now be used to compute t_i for the given c_i and the results compared against the observed t_i that were used to fit the polynomial. Note the discrepancies between the observed and predicted values due to machine round-off error.

─────────────────── **EXHIBIT 5.2.2** ───────────────────

Fitting A High-Order Polynomial To The Tape Recorder Data

```
MTB > SET C1
DATA> 1:8
DATA> END
MTB > SET C2
DATA> 2.05 4.3 6.77 9.45 12.33 15.42 18.72 22.24
DATA> END
MTB > SET C3
DATA> 1 1 1 1 1 1 1 1
DATA> END
MTB > LET C4=C1**2
MTB > LET C5=C1**3
MTB > LET C6=C1**4
MTB > LET C7=C1**5
MTB > LET C8=C1**6
MTB > LET C9=C1**7
MTB > COPY C3 C1 C4 C5 C6 C7 C8 C9 INTO M1
MTB > PRINT M1
MATRIX M1
```

1	1	1	1	1	1	1	1
1	2	4	8	16	32	64	128
1	3	9	27	81	243	729	2187
1	4	16	64	256	1024	4096	16384
1	5	25	125	625	3125	15625	78125
1	6	36	216	1296	7776	46656	279935
1	7	49	343	2401	16807	117649	823539
1	8	64	512	4096	32768	262143	2097140

```
MTB > INVERT M1 M2
MTB > COPY C2 M3
MTB > MULT M2 M3 M4
MTB > PRINT M4
MATRIX M4
```

```
 -0.13751
  2.31921
 -0.27877
  0.19118
 -0.05166
  0.00773
 -0.00060
  0.00002
```

```
MTB > NOTE THESE ARE THE COEFFICIENTS OF THE POLYNOMIAL.
```

112

```
MTB > MULT M1 M4 M5
MTB > PRINT M5
MATRIX M5

     2.0500
     4.3005
     6.7700
     9.4507
    12.3359
    15.4396
    18.7673
    22.3354


MTB > NOTE THESE ARE THE PREDICTIONS FOR TIME.
MTB > COPY M5 INTO C10
MTB > NAME C2 'TIME',C10 'PREDTIME'
MTB > PRINT C2 C10
 ROW    TIME   PREDTIME

   1    2.05     2.0500
   2    4.30     4.3005
   3    6.77     6.7700
   4    9.45     9.4507
   5   12.33    12.3359
   6   15.42    15.4396
   7   18.72    18.7673
   8   22.24    22.3354

MTB > NOTE THE DISCREPANCIES DUE TO ROUNDOFF ERROR.
```

5.2   EXERCISES

For the following data sets,

   A.   Formulate the system of equations determining the coefficients of a polynomial that passes exactly through each of the data points.

   B.   Find the coefficients of the appropriate polynomial.

   C.   Sketch a graph of the polynomial.

   D.   Decide if the polynomial represents the trend of the data.

1.

x	0	1	2	3	4	5	6	7
y	1	4.5	20	90	403	1808	8103	36316

2.

t (year)	1814	1824	1834	1844	1854	1864
P(t)	125	275	830	1200	1750	1650

113

3. length (in)	12.5	12.625	14.125	14.5	17.25	17.75	
weight (oz)	17	16.5	23	26.5	41	49	

4. bodyweight	114.5	123.5	132.5	149	165.5	182	198.5	242.5
lift	534.6	578.7	628.3	677.9	738.5	804.7	843.3	881.8

5.3   Polynomial Smoothing

Smoothing with low-order polynomials is an attempt to retain the advantages of polynomials as empirical models while at the same time reducing the tendencies of higher-order polynomials to snake and oscillate. The idea of smoothing is illustrated graphically in Figure 5-3. Rather than forcing a polynomial to pass through all the data points, a low-order polynomial is chosen and fit to the data. This choice normally results in a situation where the number of data points exceeds the number of constants necessary to determine the polynomial. Smoothing with polynomials requires two decisions:

1. Is a low-order polynomial appropriate? If so, what should the order be?

2. What are the parameters of the model according to some criterion of "best fit," such as the least-squares criterion?

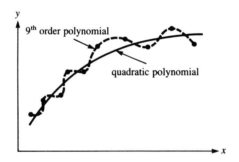

FIGURE 5-3. The quadratic function "smooths" the data since it is not required to pass through all the data points.

For example, we may decide to fit a quadratic to 10 data points, as indicated in Figure 5-3. Once that decision is made, the curve-fitting problem is the same as that addressed in Chapter 4 and the REGRESS command may be used to apply the least-squares criterion. In this section we discuss how to use Minitab to determine if a low-order polynomial is appropriate and to help us choose a suitable order of that polynomial.

Divided Difference Tables

An nth order polynomial is characterized by the properties that its nth derivative is constant and its (n+1)st derivative is identically zero. Given a set of discrete data points, we can compute their divided differences and use them to approximate the corresponding derivatives. In Table 5-2 we give the formulas for the first and second divided differences, and interpret them geometrically in Figure 5-4. For example, consider the data set in Table 5-3, where Δ^n indicates the nth divided difference. For that particular set of data the second divided difference is constant and the third divided difference is zero, which indicate that a quadratic curve would precisely pass through the 5 data points. Of course, with a real-world data set we expect the presence of measurement errors even if the underlying behavior is quadratic in nature. Thus the third divided differences are not likely to be exactly zero. The objective here is to use a divided difference table as a qualitative aid to determine if a low-order polynomial is worthy of further investigation. Ultimately, the model will be fit to the data and the goodness of fit analyzed (using techniques previously presented) to determine if the model is reasonable for predictive purposes.

TABLE 5-2

**The First And Second Divided Differences Estimate
The First And Second Derivatives, Respectively**

Data		First divided difference	Second divided difference
x_1	y_1		
		$\dfrac{y_2 - y_1}{x_2 - x_1}$	
x_2	y_2		$\dfrac{\dfrac{y_3 - y_2}{x_3 - x_2} - \dfrac{y_2 - y_1}{x_2 - x_1}}{x_3 - x_1}$
		$\dfrac{y_3 - y_2}{x_3 - x_2}$	
x_3	y_3		

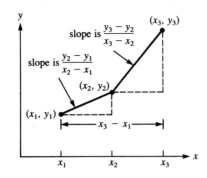

FIGURE 5-4. The second divided difference may be interpreted as the difference between the adjacent slopes (first divided differences) divided by the length of the interval over which the change has taken place.

TABLE 5-3

A Divided Difference Table For The Following Set Of Data

x_i	0	2	4	6	8
y_i	0	4	16	36	64

Data		Divided differences		
x_i	y_i	Δ	Δ^2	Δ^3
0	0			
2	4	4/2 = 2	4/4 = 1	
4	16	12/2 = 6	4/4 = 1	0/6 = 0
6	36	20/2 = 10	4/4 = 1	0/6 = 0
8	64	28/2 = 14		

$\Delta x = 6$

Constructing Divided Difference Tables Using Minitab

Minitab does not possess a single command for constructing a divided difference table. However, a program can be developed to calculate and print a divided difference table using the STORE and looping commands coupled with the CK capability of Minitab. The CK capability is a technique which allows the user to loop through columns by addressing the column number as a variable.

In Minitab a sequence of commands can be stored for future use on call. Such a sequence is called a "macro." Furthermore, we can "loop" through the macro a designated number of times using the EXECUTE command. Thus Minitab has a programming capability.

Consider the following sequence of commands and their formats:

```
MTB>STORe (in 'filename') the following Minitab commands
STOR>
    .
    .       (User's Minitab commands go here)
    .


STOR>END

MTB>EXECute commands (stored in 'filename') (K times)
```

Some basic rules applicable to these commands include:

1. If the optional K in the EXECUTE command is omitted, it is assumed to be 1. If K is 0 or negative, the commands are not executed.

2. If you omit the "filename" in the STORE command, a default filename is used. (The name varies from installation to installation--see your local documentation.)

3. If you use the EXECUTE command without specifying the optional "filename," the macro stored in the default file is executed.

4. You can nest STORE files. The depth to which you can nest the files varies from installation to installation.

5. You can use the editor on your computer to create a file of stored commands. You can also use your editor to edit a stored file that has been created in Minitab with the STORE command.

6. You cannot store an END of data command following data for a READ, SET, or INSERT command when you STORE a file of commands: the END command will be interpreted as an END to the STORE command. (However, if you form a macro in the editor you can use the END statement.)

7. Suppose we are in the process of storing the following macro which contains the SET command in order to enter data:

```
MTB>STOR 'macro'
    .
    .
    .
STOR>SET C1
STOR>1:20
STOR>END
MTB>
```

The formatted set (1:20) will not appear when the file is executed. Also, the END terminates the STORE sequence and returns control to the MTB> prompt. If you had wished to continue with additional

Minitab commands, you would simply continue with the commands and not enter the END command.

8. Do not name a column with the same name used as a store filename.

The following gives an example of the use of the STORE/END command (further examples are given in Chapter 7). In the example we are considering meteorological data in the United States.

Example 1   Meteorological Data In The United States

Suppose you have data for 12 states located in columns C1-C4. The column C1 contains an integer code for each state; C2-C4 contain rainfall, humidity, and temperature data. The following commands perform a separate regression and residual plot for the data from each state.

```
MTB>STORE 'STATES'
STOR>CHOOSE K1 in C1 C2-C4 put in C11-C14
STOR>PRINT K1
STOR>REGRESS C12 on 2 pred. C13 C14 put in C20 C21
STOR>PLOT residuals in C20 vs predictors in C21
STOR>LET K1 = K1 + 1
STOR>END
```

To execute the macro after it has been stored:

```
MTB>LET K1 = 1
MTB>EXECUTE 'STATES' 12 times
```

Additional Commands

The CK capability allows the integer portion of the column number to be replaced by a stored constant. For example:

```
MTB>LET K1 = 13
MTB>PRINT C1 - CK1
```

Since K1 = 13, C1 - C13 will be printed.

There are two additional commands that are useful for the STORE file: the NOECHO and ECHO commands. The format for these commands is:

```
MTB>NOECho the commands that follow
```

```
MTB>ECHO the commands that follow
```

When you EXECUTE a file, each command is automatically printed (echoed) along with the output. If your macro is executing correctly and you are making a large "production run," you may not want the commands to be printed with each loop. In that case you would use the NOECHO command in the stored macro file, and only the output would be printed. However, all NOTE commands will still be printed when the NOECHO command is invoked. If you want to turn the echo printing of commands back on, then use ECHO in your macro file.

We highly recommend that the NOECHO/ECHO commands not be used until you have EXECUTED the program and cleared any errors. When you have the commands printed, it helps you interpret any error messages that may occur. When the commands in your file have been verified to be correct, you can add the NOECHO and ECHO feature to your file. Normally, NOECHO will be the first command after STORE, and ECHO the last command in the store block just before the END command.

We have developed a series of nested STORE commands to construct a divided difference table. The macros 'DRIVER' and 'DIVDIV' appear in Appendix B, and they must be entered into a file that is accessible to all the intended users. Once the program in Appendix B has been entered, a divided difference table may be procured using the following sequence of commands. (It is not necessary for you to understand the commands in the program. Simply key-in the program and EXECUTE it whenever you desire a divided difference table. The examples will clarify the correct procedure.)

Program For A Divided Difference Table

```
MTB>STORE 'DDTABLE'
STOR>NOECHO
STOR>EXECUTE 'DRIVER'
STOR>EXECUTE 'DIVDIV'
STOR>LET  K400  =  200  +  K300
STOR>PRINT C1, C2, C200 - CK400
STOR>ECHO
STOR>END
```

The above program calculates all the divided differences and prints out only the divided difference table; no commands are printed. The macros 'DRIVER' and 'DIVDIV' are listed in Appendix B. We now offer several examples illustrating both the construction and the interpretation of a divided difference table.

Example 2   Vehicular Stopping Distance

In Exhibit 5.3.1 we construct a divided difference table for the data presented in Section 3.4 concerning the distance required to brake a car as a function of its velocity. An examination of the divided difference table reveals that the third divided differences are small in magnitude compared to the original data, and that negative signs have started to appear. These negative signs may indicate the presence of measurement error, or variations in the data that will not be captured by a low-order polynomial. The negative signs will have a detrimental effect on the divided differences in the remaining columns. So we may decide to use a quadratic model, reasoning that higher-order terms will not reduce the deviations significantly enough to justify their inclusion (see Giordano and Weir, op. cit., pp. 155-196.)

After a quadratic model is fit via least-squares, the residuals are determined. A residual plot aids in our evaluation of this model. It should be noted that the model differs from those previously developed in Sections 3.4 and 4.4 where the submodels were fit individually.

───────────────────── **EXHIBIT 5.3.1** ─────────────────────

**A Low-Order Polynomial Model For Vehicular
Stopping Distance**

```
MTB > RETR 'CAR'
MTB > INFO

COLUMN     NAME        COUNT
C1         SPEED          13
C2         T DIST         13

CONSTANTS USED: K1
```

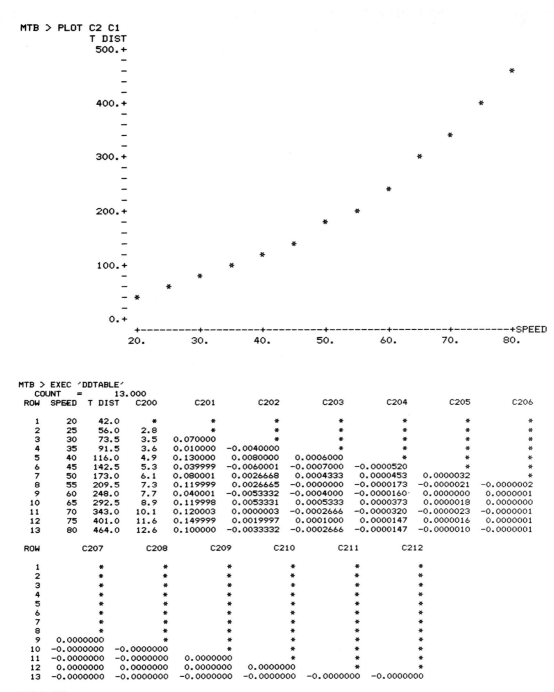

```
MTB > PLOT C2 C1
          T DIST
        500.+
            -
            -                                                                    *
            -
            -
        400.+                                                              *
            -
            -                                                        *
            -
        300.+                                                  *
            -
            -                                            *
            -
        200.+                                      *
            -                                *
            -                          *
            -                    *
        100.+              *
            -        *
            -  *
            - *
            -
          0.+
            +---------+---------+---------+---------+---------+---------+SPEED
           20.       30.       40.       50.       60.       70.       80.
```

```
MTB > EXEC 'DDTABLE'
    COUNT   =    13.000
```

ROW	SPEED	T DIST	C200	C201	C202	C203	C204	C205	C206
1	20	42.0	*	*	*	*	*	*	*
2	25	56.0	2.8	*	*	*	*	*	*
3	30	73.5	3.5	0.070000	*	*	*	*	*
4	35	91.5	3.6	0.010000	-0.0040000	*	*	*	*
5	40	116.0	4.9	0.130000	0.0080000	0.0006000	*	*	*
6	45	142.5	5.3	0.039999	-0.0060001	-0.0007000	-0.0000520	*	*
7	50	173.0	6.1	0.080001	0.0026668	0.0004333	0.0000453	0.0000032	*
8	55	209.5	7.3	0.119999	0.0026665	-0.0000000	-0.0000173	-0.0000021	-0.0000002
9	60	248.0	7.7	0.040001	-0.0053332	-0.0004000	-0.0000160	0.0000000	0.0000001
10	65	292.5	8.9	0.119998	0.0053331	0.0005333	0.0000373	0.0000018	0.0000000
11	70	343.0	10.1	0.120003	0.0000003	-0.0002666	-0.0000320	-0.0000023	-0.0000001
12	75	401.0	11.6	0.149999	0.0019997	0.0001000	0.0000147	0.0000016	0.0000001
13	80	464.0	12.6	0.100000	-0.0033332	-0.0002666	-0.0000147	-0.0000010	-0.0000001

ROW	C207	C208	C209	C210	C211	C212
1	*	*	*	*	*	*
2	*	*	*	*	*	*
3	*	*	*	*	*	*
4	*	*	*	*	*	*
5	*	*	*	*	*	*
6	*	*	*	*	*	*
7	*	*	*	*	*	*
8	*	*	*	*	*	*
9	0.0000000	*	*	*	*	*
10	-0.0000000	-0.0000000	*	*	*	*
11	-0.0000000	-0.0000000	0.0000000	*	*	*
12	0.0000000	0.0000000	0.0000000	0.0000000	*	*
13	-0.0000000	-0.0000000	-0.0000000	-0.0000000	-0.0000000	-0.0000000

```
MTB > END
MTB >
```

```
MTB > LET C3=C1**2
MTB > NAME C3 'SPEED**2'
MTB > CONS NOBRIEF
MTB > REGRESS C2 2 C1 C3;
SUBC> RESIDS IN C10.
```

THE REGRESSION EQUATION IS
T DIST = 50.1 - 1.97 SPEED + 0.0886 SPEED**2

COLUMN	COEFFICIENT	ST. DEV. OF COEF.	T-RATIO = COEF/S.D.
	50.060	6.367	7.86
SPEED	-1.9702	0.2776	-7.10
SPEED**2	0.088592	0.002738	32.35

S = 3.063

R-SQUARED =100.0 PERCENT
R-SQUARED =100.0 PERCENT, ADJUSTED FOR D.F.

ANALYSIS OF VARIANCE

DUE TO	DF	SS	MS=SS/DF
REGRESSION	2	225756	112878
RESIDUAL	10	94	9
TOTAL	12	225850	

FURTHER ANALYSIS OF VARIANCE
SS EXPLAINED BY EACH VARIABLE WHEN ENTERED IN THE ORDER GIVEN

DUE TO	DF	SS
REGRESSION	2	225756
SPEED	1	215936
SPEED**2	1	9820

ROW	SPEED	Y T DIST	PRED. Y VALUE	ST.DEV. PRED. Y	RESIDUAL	ST.RES.
1	20.0	42.000	46.093	2.201	-4.093	-1.92
2	25.0	56.000	56.176	1.605	-0.176	-0.07
3	30.0	73.500	70.688	1.251	2.812	1.01
4	35.0	91.500	89.629	1.141	1.871	0.66
5	40.0	116.000	113.000	1.182	3.000	1.06
6	45.0	142.500	140.801	1.251	1.699	0.61
7	50.0	173.000	173.032	1.281	-0.032	-0.01
8	55.0	209.500	209.691	1.251	-0.191	-0.07
9	60.0	248.000	250.781	1.182	-2.781	-0.98
10	65.0	292.500	296.299	1.141	-3.799	-1.34
11	70.0	343.000	346.248	1.251	-3.248	-1.16
12	75.0	401.000	400.626	1.605	0.374	0.14
13	80.0	464.000	459.434	2.201	4.566	2.14R

R DENOTES AN OBS. WITH A LARGE ST. RES.

DURBIN-WATSON STATISTIC = 0.74

```
MTB > NAME C10 'RES'
MTB > PLOT C10 C1
```

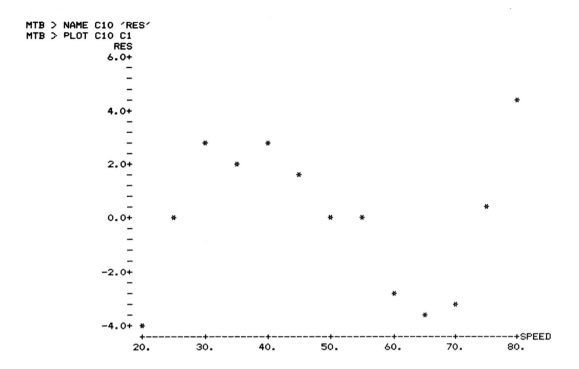

```
MTB > LET C5=50.06

MTB > SET C5
DATA> 1:80
DATA> END
MTB > LET C6=50.06-1.9702*C5+.088592*C5**2
MTB > WIDTH 80,50
MTB > MPLOT C2 C1 C6 C5
```

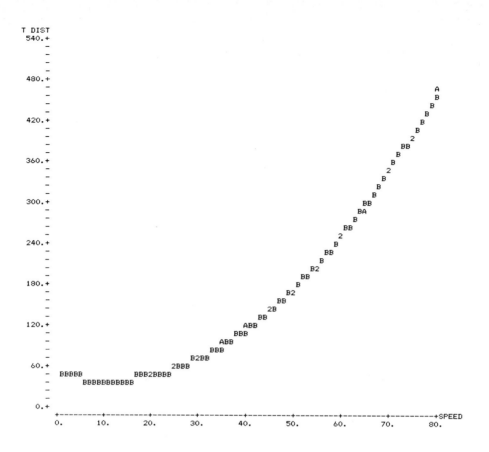

Example 3   Growth Of A Yeast Culture

In Exhibit 5.3.2 data are presented which represent the population of yeast cells in a culture as a function of time measured in hours. Inspection of the scatterplot indicates that a simple one-term model will not suffice. Examination of the divided difference table reveals that the first 9 second divided differences are positive and the last 10 are negative. This change in sign of the second derivative signifies a change in concavity, which is supported visually in the scatterplot. The lowest order polynomial to exhibit a change in concavity is a cubic, which is the curve we decide to try.

After fitting a third order polynomial to the given data, a residual plot aids in our evaluation of the model. A definite trend is seen in the residual plot which signals that we might want to look for another model type (see Giordano and Weir, <u>op. cit.</u>, pp. 310-313).

A Low-Order Polynomial Model For The Growth
Of Yeast In A Culture

```
MTB > RETR 'YEAST'
MTB > INFO

COLUMN     NAME       COUNT
C1         TIME          19
C2         CULTS         19

CONSTANTS USED: NONE
```

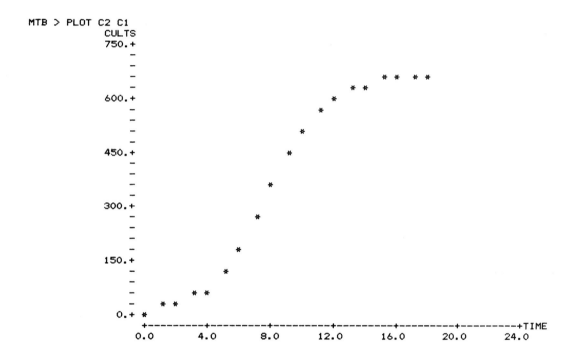

```
MTB > PLOT C2 C1
      CULTS
     750.+
         -
         -
         -
         -                                              *  *    *  *
         -                                        *  *
     600.+                                   *
         -                                *
         -                           *
         -
         -                     *
     450.+                  *
         -
         -
         -             *
         -
     300.+        *
         -
         -
         -
         -    *
     150.+
         -   *
         -
         - *  *
         -
       0.+ *
         +---------+---------+---------+---------+---------+---------+TIME
         0.0       4.0       8.0      12.0      16.0      20.0      24.0
```

```
MTB > EXEC 'DDTABLE'
   COUNT   =      19.000
   ROW  TIME   CULTS      C200       C201       C202       C203       C204       C205        C206

    1     0    9.6         *          *          *          *          *          *           *
    2     1   18.3      8.7000        *          *          *          *          *           *
    3     2   29.0     10.7000     1.0000        *          *          *          *           *
    4     3   47.2     18.2000     3.7500     0.91667       *          *          *           *
    5     4   71.1     23.9000     2.8500    -0.30000   -0.30417       *          *           *
    6     5  119.1     48.0000    12.0500     3.06667    0.84167    0.229167       *           *
    7     6  174.6     55.5000     3.7500    -2.76667   -1.45833   -0.460000  -0.114861        *
    8     7  257.3     82.7000    13.6000     3.28334    1.51250    0.594167   0.175695    0.0415080
    9     8  350.7     93.4000     5.3500    -2.75001   -1.50834   -0.604168  -0.199723   -0.0536310
   10     9  441.0     90.3000    -1.5500    -2.29998    0.11251    0.324169   0.154723    0.0506351
   11    10  513.3     72.2999    -9.0001    -2.48337   -0.04585   -0.031671  -0.059307   -0.0305757
   12    11  559.7     46.4000   -12.9500    -1.31663    0.29168    0.067506   0.016530    0.0108338
   13    12  594.8     35.1000    -5.6500     2.43331    0.93748    0.129160   0.010276   -0.0008934
   14    13  629.4     34.6000    -0.2500     1.80001   -0.15833   -0.219162  -0.058054   -0.0097613
   15    14  640.8     11.4000   -11.6000    -3.78333   -1.39583   -0.247502  -0.004723    0.0076186
   16    15  651.1     10.3000    -0.5500     3.68333    1.86666    0.652499   0.150000    0.0221033
   17    16  655.9      4.7999    -2.7501    -0.73336   -1.10417   -0.594167  -0.207778   -0.0511111
   18    17  659.6      3.7001    -0.5499     0.73338    0.36668    0.294171   0.148056    0.0508334
   19    18  661.8      2.2000    -0.7501    -0.06671   -0.20002   -0.113341  -0.067919   -0.0308536
```

```
MTB > REGRESS C2 3 C1 C3 C4;
SUBC> RESIDS IN C11.
* NOTE *     T**2 IS HIGHLY CORRELATED WITH OTHER PREDICTOR VARIABLES
* NOTE *     T**3 IS HIGHLY CORRELATED WITH OTHER PREDICTOR VARIABLES

THE REGRESSION EQUATION IS
CULTS = 8.2 - 13.2 TIME + 10.1 T**2 - 0.417 T**3


                                    ST. DEV.      T-RATIO =
   COLUMN      COEFFICIENT          OF COEF.      COEF/S.D.
                  8.21               19.66          0.42
   TIME         -13.157              9.721         -1.35
   T**2          10.131              1.275          7.94
   T**3         -0.41679            0.04652        -8.96

S = 25.78

R-SQUARED = 99.2 PERCENT
R-SQUARED = 99.1 PERCENT, ADJUSTED FOR D.F.
```

ANALYSIS OF VARIANCE

DUE TO	DF	SS	MS=SS/DF
REGRESSION	3	1249977	416659
RESIDUAL	15	9965	664
TOTAL	18	1259943	

FURTHER ANALYSIS OF VARIANCE
SS EXPLAINED BY EACH VARIABLE WHEN ENTERED IN THE ORDER GIVEN

DUE TO	DF	SS
REGRESSION	3	1249977
TIME	1	1179551
T**2	1	17100
T**3	1	53326

ROW	TIME	Y CULTS	PRED. Y VALUE	ST.DEV. PRED. Y	RESIDUAL	ST.RES.
1	0.0	9.60	8.21	19.66	1.39	0.08
2	1.0	18.30	4.77	13.44	13.53	0.62
3	2.0	29.00	19.09	10.59	9.91	0.42
4	3.0	47.20	48.66	10.17	-1.46	-0.06
5	4.0	71.10	91.00	10.50	-19.90	-0.85
6	5.0	119.10	143.60	10.62	-24.50	-1.04
7	6.0	174.60	203.95	10.30	-29.35	-1.24
8	7.0	257.30	269.56	9.71	-12.26	-0.51
9	8.0	350.70	337.92	9.13	12.78	0.53
10	9.0	441.00	406.54	8.89	34.46	1.42
11	10.0	513.30	472.91	9.13	40.39	1.68
12	11.0	559.70	534.54	9.71	25.16	1.05
13	12.0	594.80	588.92	10.30	5.88	0.25
14	13.0	629.40	633.56	10.62	-4.16	-0.18
15	14.0	640.80	665.94	10.50	-25.14	-1.07
16	15.0	651.10	683.58	10.17	-32.48	-1.37
17	16.0	655.90	683.96	10.59	-28.06	-1.19
18	17.0	659.60	664.60	13.44	-5.00	-0.23
19	18.0	661.80	622.98	19.66	38.81	2.33R

R DENOTES AN OBS. WITH A LARGE ST. RES.

DURBIN-WATSON STATISTIC = 0.58

```
MTB > NAME C11 'RES'
MTB > PLOT C11 C1
```
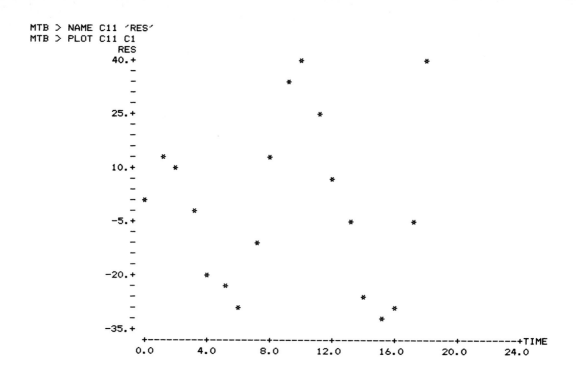
```
MTB > SET C3
DATA> 0:19/.5
DATA> END
MTB > LET C4=8.21-13.157*C3+10.131*C3**2-.41679*C3**3
MTB > WIDTH 80,50
MTB > MPLOT C2 C1 C4 C3
```

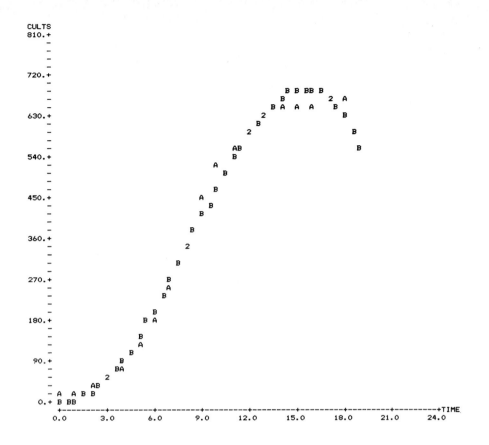

5.3 EXERCISES

For the data sets in Problems 1 through 4, use Minitab to construct
the divided difference table. What conclusions can you draw about
the data? Would you model the data with a low order polynomial?
If so, what order would you choose?

1.

x	0	1	2	3	4	5	6	7
y	2	8	24	56	110	192	308	464

2.

x	0	1	2	3	4	5	6	7
y	23	48	73	98	123	148	173	198

3.

x	0	1	2	3	4	5	6	7
y	7	15	33	61	99	147	205	273

4.

x	0	1	2	3	4	5	6	7
y	1	4.5	20	90	403	1803	8103	36316

129

In problems 5-11 construct a scatterplot of the data. Is there any
trend?   Are any data points "outliers?"   Construct a divided
difference table. Is smoothing with a low-order polynomial appropriate? If so, choose an appropriate low-order polynomial and fit
the data using the least squares criterion of best fit.   Analyze
the fit by examining appropriate indicators and plotting the model,
the data points, and the residuals.

5.

x	46	49	51	52	54	56	57	58	59	60	61
y	40	50	55	63	72	70	77	73	90	93	96

x	62	63	64	66	67	68	71	72
y	88	99	110	113	120	127	137	132

where x is the Fahrenheit temperature and y is the number of
times a cricket chirps in one minute.

6.

x	17	19	20	22	23	25	31	32	33	36	37
y	19	25	32	51	57	71	141	123	187	192	205

x	38	39	41
y	252	248	294

where x is the diameter of a ponderosa pine measured at breast
height and y is a measure of volume--number of board feet divided
by 10.

7.   The following data represent the population of the United
States from 1790 to 1980.

Year	Observed Population
1790	3,929,000
1800	5,308,000
1810	7,240,000
1820	9,638,000
1830	12,866,000
1840	17,069,000
1850	23,192,000
1860	31,443,000
1870	38,558,000
1880	50,156,000
1890	62,948,000
1900	75,995,000
1910	91,972,000
1920	105,711,000
1930	122,755,000
1940	131,669,000
1950	150,696,000
1960	179,323,000
1970	203,212,000
1980	226,505,000

8. The following data were obtained for the growth of a sheep population introduced into a new environment on the island of Tasmania. (Adapted from J. Davidson, "On the Growth of the Sheep Population in Tasmania," Trans. Roy. Soc. S. Australia 62 (1938): 343-346.)

t (year)	1814	1824	1834	1844	1854	1864
P(t)	125	275	830	1200	1750	1650

9. The following data represent the "pace of life." Here P is the population and V the mean velocity in feet per second over a 50-ft course.

P	365	2500	5491	14000	23700	49375	70700	78200
V	2.76	2.27	3.31	3.70	3.27	4.90	4.31	3.85

P	138000	304500	341948	867023	1092759	1340000	2602000
V	4.39	4.42	4.81	5.21	5.88	5.62	5.05

10. The following data represent the length of a fish (bass) and its weight.

length (in.)	12.5	12.625	14.125	14.5	17.25	17.75
weight (oz)	17.0	16.50	23.00	26.5	41.00	49.00

11. The following data represent the weight-lifting results from the 1976 Olympics.

Bodyweight class (lb)		Total winning lifts (lb)		
	Max. weight	Snatch	Jerk	Total weight
Flyweight	114.5	231.5	303.1	534.6
Bantamweight	123.5	259.0	319.7	578.7
Featherweight	132.5	275.6	352.7	628.3
Lightweight	149.0	297.6	380.3	677.9
Middleweight	165.5	319.7	418.9	738.5
Light-heavyweight	182.0	358.3	446.4	804.7
Middle-heavyweight	198.5	374.8	468.5	843.3
Heavyweight	242.5	385.8	496.0	881.8

5.4 Cubic Splines

The use of polynomials in constructing empirical models that capture the trend of the data is appealing because polynomials are easy to integrate and differentiate. However, high-order polynomials tend to oscillate near the endpoints of the data interval. Although smoothing with a low-order polynomial lessens these effects, unless the data are essentially quadratic or cubic in nature, a low-order polynomial may give a relatively poor fit somewhere over the range of data. In this section we introduce cubic spline interpolation as an alternative method for constructing empirical models. By using different cubic polynomials between successive pairs of data points and by connecting the cubics together in a smooth fashion, we can capture the trend of the data, regardless of the underlying relationships. Simultaneously we will reduce the tendency toward oscillation and the sensitivity of the coefficients to changes in the data.

In Figure 5-5 the cubic spline model consists of the two segments $S_1(x)$ and $S_2(x)$. We define each separate spline function on its respective interval as follows:

$$S_1(x) = a_1 + b_1x + c_1x^2 + d_1x^3 \quad \text{for x in } [x_1,x_2)$$

$$S_2(x) = a_2 + b_2x + c_2x^2 + d_2x^3 \quad \text{for x in } [x_2,x_3] \; .$$

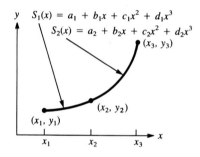

FIGURE 5-5 A cubic spline model is a continuous function with continuous first and second derivatives consisting of cubic polynomial segments.

We are going to need the first and second derivatives of each spline segment, so let's compute them as well:

$$S_1'(x) = b_1 + 2c_1x + 3d_1x^2 \quad \text{for x in } [x_1,x_2)$$

$$S_1''(x) = 2c_1 + 6d_1x \quad \text{for x in } [x_1,x_2)$$

$$S_2'(x) = b_2 + 2c_2x + 3d_2x^2 \quad \text{for x in } [x_2,x_3]$$

$$S_2''(x) = 2c_2 + 6d_2x \quad \text{for x in } [x_2,x_3]$$

The smoothness across the cubic spline segments is obtained by requiring that the first two derivatives of the adjacent segments match (i.e., have equal values) at the interior nodes where they meet. Thus, in our example, $S_1'(x_2) = S_2'(x_2)$ and $S_1''(x_2) = S_2''(x_2)$. In general, we construct a cubic spline model by applying the following conditions:

> 1. Require each spline segment to pass through the two data points specifying the interval over which the segment is defined.
>
> 2. Match the 1st and 2nd derivatives of adjacent spline segments at the interior nodes where they meet.
>
> 3. Apply an "end condition" to the extreme left and right spline segments--the exterior segments. The two popular end conditions give the "natural" and "clamped" spline types. We illustrate both of these splines in our next example.

Let's construct a cubic spline for the data set:

x_i	1	2	3
y_i	5	8	25

The first condition requires that the following equations be satisfied:

$$a_1 + 1b_1 + 1c_1 + 1d_1 = 5$$

$$a_1 + 2b_1 + 2^2c_1 + 2^3d_1 = 8$$

$$a_2 + 2b_2 + 2^2c_2 + 2^3d_2 = 8$$

$$a_2 + 3b_2 + 3^2c_2 + 3^3d_2 = 25$$

The second condition gives the equations:

$$b_1 + 2c_1(2) + 3d_1(2^2) = b_2 + 2c_2(2) + 3d_2(2^2)$$

$$2c_1 + 6d_1(2) = 2c_2 + 6d_2(2)$$

The Natural Spline

For the third condition, the natural spline condition requires that the 2nd derivatives for the exterior splines equal zero at the exterior endpoints. For our example this requirement yields $S_1''(x_1) = 0$ and $S_2''(x_3) = 0$, or

$$2c_1 + 6d_21 = 0$$

$$2c_2 + 6d_23 = 0$$

Summarizing these equations and writing them in matrix form gives:

133

$$
\begin{bmatrix}
1 & 1 & 1 & 1 & 0 & 0 & 0 & 0 \\
1 & 2 & 4 & 8 & 0 & 0 & 0 & 0 \\
0 & 0 & 0 & 0 & 1 & 2 & 4 & 8 \\
0 & 0 & 0 & 0 & 1 & 3 & 9 & 27 \\
0 & 1 & 4 & 12 & 0 & -1 & -4 & -12 \\
0 & 0 & 2 & 12 & 0 & 0 & -2 & -12 \\
0 & 0 & 2 & 6 & 0 & 0 & 0 & 0 \\
0 & 0 & 0 & 0 & 0 & 0 & 2 & 18
\end{bmatrix}
\begin{bmatrix}
a_1 \\ b_1 \\ c_1 \\ d_1 \\ a_2 \\ b_2 \\ c_2 \\ d_2
\end{bmatrix}
=
\begin{bmatrix}
5 \\ 8 \\ 8 \\ 25 \\ 0 \\ 0 \\ 0 \\ 0
\end{bmatrix}
$$

The above matrix equation can be solved conveniently in Minitab using the READ, INVERT and MULTIPLY commands presented in Section 5.2.

In Exhibit 5.4.1 the above system is solved to yield the following empirical model:

Interval	Model
$1 \leq x < 2$	$S_1(x) = 2 + 10x - 10.5x^2 + 3.5x^3$
$2 \leq x \leq 3$	$S_2(x) = 58 - 74x + 31.5x^2 - 3.5x^3$

The spline segments are sketched in the exhibit. Note how the graphs of the different cubic segments fit together to form a single smooth interpolating curve joining the data points.

───────── EXHIBIT 5.4.1 ─────────

Constructing A Natural Cubic Spline

```
MTB > READ 8 8 MATRIX M1
DATA> 1 1 1 1 0 0 0 0
DATA> 1 2 4 8 0 0 0 0
DATA> 0 0 0 0 1 2 4 8
DATA> 0 0 0 0 1 3 9 27
DATA> 0 1 4 12 0 -1 -4 -12
DATA> 0 0 2 12 0 0 -2 -12
DATA> 0 0 2 6 0 0 0 0
DATA> 0 0 0 0 0 0 2 18
     8 ROWS READ
```

```
MTB > READ 8 1 M2
DATA> 5
DATA> 8
DATA> 8
DATA> 25
DATA> 0
DATA> 0
DATA> 0
DATA> 0
     8 ROWS READ
MTB > INVERT M1 M3
MTB > MULT M3 M2 M4
MTB > PRINT M4
MATRIX M4

   2.0
  10.0
 -10.5
   3.5
  58.0
 -74.0
  31.5
  -3.5


MTB >
```

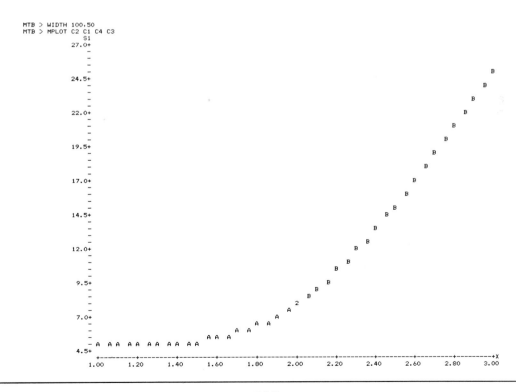

The Clamped Spline

If the values of the first derivatives at the exterior end-points are known, the first derivatives of the exterior segments can be required to match the known values. In our previous example, suppose the derivatives at the exterior endpoints are known to have the values $f'(x_1)$ and $f'(x_3)$. Mathematically, the clamped spline matching requirement yields the equations:

$$S_1'(x_1) = b_1 + 2c_1x_1 + 3d_1x_1{}^2 = f'(x_1)$$

$$S_2'(x_3) = b_2 + 2c_2x_3 + 3d_2x_3{}^2 = f'(x_3).$$

The construction of cubic splines for more than three data points proceeds in the same manner. That is, each spline segment is forced to pass through the endpoints of the interval over which it is defined, the first and second derivatives of adjacent splines are forced to match at the interior data points where they meet, and a set of end conditions (for example, the natural or clamped spline conditions) are applied at the two exterior data points. However, as we have previously observed, large systems of linear equations tend to be difficult to solve with great numerical precision. The procedure we have described does not lend itself to computationally efficient and accurate computer algorithms. Nevertheless, our approach does facilitate your understanding of the basic concepts underlying cubic spline interpolation. For a computationally efficient algorithm, see Richard L. Burden and J. Douglas Faires, <u>Numerical Analysis</u>, Third edition, Prindle, Weber & Schmidt, Boston, Mass., 1985, pp. 122 and 124.

5.4   EXERCISES

For each of the following data sets, write a system of equations to determine the coefficients of the natural cubic splines passing through the given points. Using Minitab, solve the system of equations and graph the splines.

1. a.

x	2	4	7
y	2	8	12

   b.

x	3	4	6
y	10	15	35

   c.

x	0	1	2
y	0	10	30

   d.

x	0	2	4
y	5	10	40

2.   The following data sets appear in 5.3 Exercises, page 131:

   a.   Exercise 8

   b.   Exercise 9

   c.   Exercise 10

   d.   Exercise 11

5.4   PROJECTS

For each scenario below:

   a.   Obtain a scatterplot of the original data and discuss any trends.   Are there any outliers?

   b.   Obtain the following models, if appropriate:

       i.   A simple one-term model

       ii.   A low-order polynomial (if suggested by a divided difference table)

       iii.   A cubic spline model

   c.   Appraise the given data set. How accurate do you think the data are?   How confident are you that the data are an adequate representation of what you are trying to model?

   d.   Pick a model for answering the questions posed in each scenario.   Justify your choice.

SCENARIO 1--THE POSTAGE STAMP

In 1981 the American public was subjected to two postage rate increases in the same year, for the first time in history.   This increase leads one to speculate what future increases might be expected in our postage rates.   The following data give the price of a 1st class postage stamp for the past century:

DATE	COST	
1885	.02	
1917	.03	
1919	.02	
1932	.03	
1958	.04	
1963	.05	
1968	.06	
1971	.08	
1974	.10	
1975	.13	(temporary)
1976	.13	
1978	.15	
1981	.18	March 22
1981	.20	November 1
1985	.22	

1. What will the price of a postage stamp be in the year 2001?

2. When will the price of a stamp reach $1.00?

3. When will the next price increase take place, and how much do you think it will be?

SCENARIO 2--MEDICARE, THE RISING SHARE PAID BY THE PATIENT

The following article was extracted from The New York Times, 8 March 1985:

Because of Federal efforts to reduce the soaring costs of Medicare, elderly beneficiaries are seeing sharp increases in their out-of-pocket expenses for a stay in the hospital. Legislators and lobbyists for the elderly were caught unaware by the increases, which were caused by a change in the Government system of figuring hospital reimbursements.

President Reagan has asked Congress to cut $19 billion from the Medicare budget over a 3-year period. Without reductions, the Medicare budget, now running at $66 billion a year, is projected to be $70 billion for the fiscal year beginning 1 October 1985. Of that amount, $48 billion will go for hospital care.

Since the Government established the new hospital payment system in 1983 as part of its continuing effort to control costs, the average length of a hospital stay for a Medicare patient has been shortened by 2 days. When the Medicare health and hospital program was established in 1965, the cost of one day's hospitalization was $40 and the average length of stay for a Medicare patient was 14 days. In 1984 the cuts caused the cost to rise to $360 a day with an average length of stay reduced from 9.5 to 7.5 days.

By compressing the period of hospitalization into fewer days, the average daily cost of hospitalization rose steeply thereby causing a sharp rise in the hospital patient's payment for each stay. Such payments were estimated by officials to total $4 billion in 1984. The table below gives a Medicare patient's share of the average daily cost of a stay in the hospital.

YEAR	COST PER DAY PAID BY PATIENT
1966	40
1967	40
1968	41
1969	49.9
1970	51
1971	65
1972	75
1973	77
1974	80
1975	95
1976	110
1977	125
1978	145
1979	165
1980	180
1981	210
1982	260
1983	310
1984	360
1985	400

1. Predict the cost per day in 1986 and 1990.

2. When will the cost per day reach $900?

6

DIMENSIONAL ANALYSIS AND SIMILITUDE

```
MATRIX ENTRY:
        READ
        COPY

MATRIX OPERATIONS:
        INVERT
        TRANSPOSE
        ADD
        SUBTRACT
        MULTIPLICATION
```

In the process of constructing a mathematical model we identify the variables influencing the behavior and attempt to determine a relationship among those retained for consideration. In the case of a single dependent variable y this procedure gives rise to some unknown function

$$y = f(x_1, x_2, \ldots, x_n)$$

where the x_i measure the various factors influencing the phenomenon under investigation. In many cases, especially for those models designed to predict some physical phenomenon, we may find it difficult or impossible to construct a solvable or tractable "explicative" model due to the inherent complexity of the problem. In certain instances we might conduct a series of experiments to see how the dependent variable y is related to various values of the independent variables. In such cases we usually prepare a figure or table and apply an appropriate curve-fitting or interpolation method that can be used to predict the value of y for suitable ranges of the independent variable(s), as was done in the previous chapter. Thus we fit an empirical model to the collected experimental data. Since the experiments which must be conducted are usually both costly and time consuming, we would like to reduce their number.

6.1   The Process Of Dimensional Analysis

Dimensional analysis is a method for reducing significantly (exponentially) the amount of experimental data that must be collected when we are modeling physical behavior. It is based on the premise that physical quantities have dimensions and that

140

physical laws are not altered by changing the units measuring dimensions. Thus the phenomenon under investigation can be described by a dimensionally correct equation among the variables. For example, if the dimensions of the variables under consideration are measured in the MLT (Mass, Length, Time) system, the process of dimensional analysis ensures that a dimensionally correct equation results by requiring the terms relating the variables be consistent in the dimensions M, L and T. In the process three conditions are placed on the variables under consideration and a system of linear algebraic equations must be resolved. The outcome of a dimensional analysis, in general, will be a reduction by three in the number of arguments of the function that ultimately must be approximated empirically. For a detailed presentation of dimensional analysis see Giordano and Weir, op. cit., pp. 216-237. Here we briefly overview the process and present a simple example before showing you how to implement the procedure using Minitab.

The fundamental result in dimensional analysis, which provides for the construction of dimensionally correct equations, is Buckingham's Theorem. This theorem gives a necessary and sufficient condition for an equation to be dimensionally homogeneous; that is, for the equation to be true regardless of the system of units in which the variables are measured. The theorem is stated as follows:

Buckingham's Theorem:

> An equation is dimensionally homogeneous if and only if it can be put into the form:
>
> $$f(\Pi_1, \Pi_2, \ldots, \Pi_n) = 0$$
>
> where f is some function of n arguments and $\{\Pi_1, \Pi_2, \ldots, \Pi_n\}$ is a complete set of dimensionless products.

We will present several examples illustrating what is meant by a "complete set of dimensionless products" in the applications of Buckingham's Theorem below. As you will see, the power of a dimensional analysis process is its reduction in the number of arguments of the original function. For example, suppose we hypothesize that the variable x_1 is dependent upon n+2 other variables $x_2, x_3, \ldots, x_{n+3}$. Symbolically, we write the equation

$$h(x_1, x_2, \ldots, x_{n+3}) = 0.$$

A dimensional analysis in the MLT System will yield, in general, a new equation

$$f(\Pi_1, \Pi_2, \ldots, \Pi_n) = 0$$

where the Π_i are dimensionless products of the x_i. Note that after a dimensional analysis we have an arbitrary function f to approximate by fitting an empirical model to collected experimental data. However, in the MLT system that function will have 3 fewer arguments. The reduction in the number of arguments from $n+3$ to n reduces exponentially the number of experiments that must be conducted. We now illustrate the process.

Example A Simple Undamped Pendulum

Consider the simple pendulum illustrated in Figure 6.1. Suppose for a particular pendulum we can obtain a reasonable approximation for the period of the pendulum by neglecting friction and drag forces. Thus, we choose to consider the variables represented in the following equation

$$t = f(m,g,r,\theta)$$

where

    t: period of the pendulum
    m: mass of the pendulum
    g: acceleration due to gravity
    r: length of the pendulum
    θ: initial angle of displacement

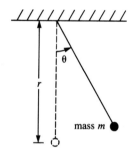

FIGURE 6.1 A simple pendulum.

The dimension of each of these variables is tabulated as follows:

Variable	m	g	t	r	θ
Dimension	M	LT^{-2}	T	L	$M^0L^0T^0$

A <u>product</u> of these variables is an expression of the form

$$m^a g^b t^c r^d \theta^e$$

and the dimension of this product is then

$$(M)^a (LT^{-2})^b (T)^c (L)^d (M^0L^0T^0)^e.$$

Now, requiring that the above product be <u>dimensionless</u> means the following equation must be satisfied:

$$M^a(LT^{-2})^b T^c L^d (M^0 L^0 T^0)^e = M^0 L^0 T^0.$$

Using the laws of exponents to collect like terms gives

$$M^{a+0e} L^{b+d+0e} T^{c-2b+0e} = M^0 L^0 T^0,$$

and equating the exponents of like terms on both sides of this latter equation yields three algebraic equations, one for the exponent of each term M, L, and T respectively:

$$M: \quad a \qquad\qquad\qquad + 0e = 0$$

$$L: \qquad\quad b \qquad + d + 0e = 0$$

$$T: \quad - 2b + c \qquad\quad + 0e = 0$$

The above system involves 5 unknowns in 3 independent equations (i.e., the rank of the system is 3). Thus, 5-3=2 of the unknowns are arbitrary variables. If we choose b and e to be arbitrary, one dimensionless product is obtained by setting b=1 and e=0 which yields the solution a=0, b=1, c=2, d=-1, e=0 and the corresponding dimensionless product $\Pi_1 = m^0 g^1 t^2 r^{-1} \theta^0$, or gt^2/r. A second dimensionless product can be obtained by setting b=0 and e=1, yielding the solution a=b=c=d=0, e=1 corresponding to the product $\Pi_2 = m^0 g^0 t^0 r^0 \theta^1$, or simply θ. Thus the two dimensionless products are $\Pi_1 = gt^2/r$ and $\Pi_2 = \theta$. Application of Buckingham's Theorem then guarantees the existence of a function H with

$$H(gt^2/r, \theta) = 0,$$

or, solving for the period t,

$$t = (r/g)^{1/2} h(\theta).$$

Note that the arbitrary function h now has one argument. Prior to the dimensional analysis the original unknown function f had 4 arguments:

$$t = f(m, g, r, \theta)$$

Since $t(g/r)^{1/2}$ is a function of θ, we can collect data for various values of $(\theta, t(g/r)^{1/2})$ and approximate the function empirically, as suggested in Figure 6.2.

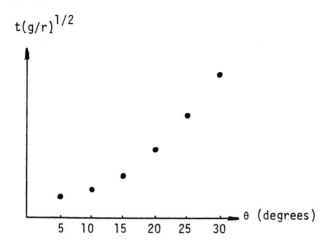

FIGURE 6.2 An empirical model would be fit to a plot of $t(g/r)^{1/2}$ vs θ.

To apply dimensional analysis we must, in general, resolve a system of linear algebraic equations containing more unknowns than equations. This task can easily be performed using Minitab. Let's review the solution procedure for such systems before presenting the necessary Minitab commands. Then, after we present the commands, we will demonstrate their use in several applications requiring a dimensional analysis.

Linear Systems Having More Unknowns Than Equations

Consider the following system containing the same number of linear equations as unknowns:

$$x_1 - x_2 - x_3 = 0$$

$$5x_1 + 20x_3 = 50$$

$$10x_2 - 20x_3 = 30$$

Elementary algebraic elimination of all but one unknown from each equation yields the solution

$$x_1 = 6$$

$$x_2 = 5$$

$$x_3 = 1$$

The necessary operations for the elimination of the unknowns can be performed on the augmented matrix:

$$\begin{bmatrix} 1 & -1 & -1 & \vdots & 0 \\ 5 & 0 & 20 & \vdots & 50 \\ 0 & 10 & -20 & \vdots & 30 \end{bmatrix} = \begin{bmatrix} 1 & 0 & 0 & \vdots & 6 \\ 0 & 1 & 0 & \vdots & 5 \\ 0 & 0 & 1 & \vdots & 1 \end{bmatrix}$$

Another technique sometimes used is to solve the matrix equation

$$AX = B$$

by premultiplying both sides by A^{-1}, the inverse of A. Here A represents the matrix of coefficients, X the column of unknowns, and B the right-hand side of the system of equations. Thus,

$$A^{-1}AX = A^{-1}B$$

yielding the solution

$$X = A^{-1}B.$$

In the above example we have,

$$\begin{bmatrix} 4/7 & 3/35 & 4/70 \\ -2/7 & 2/35 & 5/70 \\ -1/7 & 1/35 & -1/70 \end{bmatrix} \begin{bmatrix} 1 & -1 & -1 \\ 5 & 0 & 20 \\ 0 & 10 & -20 \end{bmatrix} \begin{bmatrix} x_1 \\ x_2 \\ x_3 \end{bmatrix} = \begin{bmatrix} 4/7 & 3/35 & 4/70 \\ -2/7 & 2/35 & 5/70 \\ -1/7 & 1/35 & -1/70 \end{bmatrix} \begin{bmatrix} 0 \\ 50 \\ 30 \end{bmatrix}$$

which yields

$$\begin{bmatrix} x_1 \\ x_2 \\ x_3 \end{bmatrix} = \begin{bmatrix} 6 \\ 5 \\ 1 \end{bmatrix}$$

The same result can be obtained by multiplying the entire augmented coefficient matrix by A^{-1}:

$$\begin{bmatrix} 4/7 & 3/35 & 4/70 \\ -2/7 & 2/35 & 5/70 \\ -1/7 & 1/35 & -1/70 \end{bmatrix} \begin{matrix} x_1 \ \ x_2 \ \ x_3 \\ \begin{bmatrix} 1 & -1 & -1 & \vdots & 0 \\ 5 & 0 & 20 & \vdots & 50 \\ 0 & 10 & -20 & \vdots & 30 \end{bmatrix} \end{matrix} = \begin{matrix} x_1 \ \ x_2 \ \ x_3 \\ \begin{bmatrix} 1 & 0 & 0 & \vdots & 6 \\ 0 & 1 & 0 & \vdots & 5 \\ 0 & 0 & 1 & \vdots & 1 \end{bmatrix} \end{matrix}$$

145

Note that the solution $x_1 = 6$, $x_2 = 5$, $x_3 = 1$ can be read directly from the final augmented matrix. This interpretation is made even more convenient if we use headers to label the columns corresponding to the coefficients of the respective variables, as we have demonstrated above.

To solve the system of equations resulting from the application of dimensional analysis we will likewise use an augmented matrix and premultiply by an inverse matrix. However, our system will always have 0's on the right-hand side corresponding to zero exponents for M, L, and T. Normally, the system will have more unknowns than equations. For example consider the system of equations that resulted from the pendulum problem presented previously:

$$a \qquad\qquad\qquad + 0e = 0$$

$$b \qquad + d + 0e = 0$$

$$-2b + c \qquad + 0e = 0$$

We can solve this system for 3 of the variables (the dependent variables) in terms of the remaining two (the arbitrary variables). We discuss restrictions on choosing the dependent and arbitrary variables further on. For now, let's suppose we decide to solve for the variables a, c, and d in terms of b and e, yielding the equations

$$a \qquad\qquad = 0 \qquad - 0e$$

$$d = 0 - b - 0e$$

$$c \qquad = 0 + 2b - 0e$$

Assigning the values b=1, e=0 to the arbitrary variables yields one solution a=0, b=1, c=2, d=-1, e=0. A second independent solution can be obtained by assigning the values b=0, e=1 to yield the solution a=b=c=d=0, e=1.

Next let's consider solving the above system using an augmented matrix and matrix multiplication. Note that the right-hand side consists of all 0's. We begin by transposing the arbitrary variables to the right-hand side by subtraction. The augmented matrix of coefficients is (with the columns labeled for convenience)

$$
\begin{array}{ccccc}
\mathbf{a} & \mathbf{c} & \mathbf{d} & \mathbf{b} & \mathbf{e} \\
\end{array}
$$

$$
\left[
\begin{array}{ccc|cc}
1 & 0 & 0 & 0 & 0 \\
0 & 0 & 1 & -1 & 0 \\
0 & 1 & 0 & 2 & 0 \\
\end{array}
\right]
$$

$$
\begin{array}{cc}
\textbf{MATRIX A} & \textbf{MATRIX B}
\end{array}
$$

146

Note that when convenient values (i.e., 0's and 1's) are assigned to the arbitrary variables, the values for the dependent variables are then determined by simple inspection: merely coverup all columns corresponding to the arbitrary variable(s) that are assigned the value 0, and the remaining column reveals the associated solution. All the arbitrary variables are assigned the value 0 except for one, which is assigned the value 1, and we simply progress through each arbitrary variable in succession to find a complete set of independent solutions. In effect we just move from column to column in the submatrix B.

Although the solution is obvious in the above case, we generally will solve the system by multiplying the entire augmented matrix by the inverse of the matrix of coefficients associated with the dependent variables (labeled matrix A in our above example). For instance, the product

$$
\begin{bmatrix} 1 & 0 & 0 \\ 0 & 0 & 1 \\ 0 & 1 & 0 \end{bmatrix}
\begin{array}{ccccc} a & c & d & b & e \end{array}
\begin{bmatrix} 1 & 0 & 0 & | & 0 & 0 \\ 0 & 0 & 1 & | & -1 & 0 \\ 0 & 1 & 0 & | & 2 & 0 \end{bmatrix}
$$

yields the matrix

$$
\begin{array}{ccccc} a & c & d & b & e \end{array}
\begin{bmatrix} 1 & 0 & 0 & | & 0 & 0 \\ 0 & 1 & 0 & | & 2 & 0 \\ 0 & 0 & 1 & | & -1 & 0 \end{bmatrix}
$$

as before. If we assign the values b=1, e=0 we obtain the first solution by covering the column corresponding to the variable e; the solution corresponding to the assignment b=0, e=1 is obtained by covering the column corresponding to the column b.

SUMMARY OF THE SOLUTION PROCEDURE

The procedure illustrated above is accomplished by the following four steps.

STEP 1: Determine the number of arbitrary variables. This will be the number of variables minus the rank of the system. For the MLT system, the procedure requires that the rank of the system be 3. We illustrate the solution of a system with rank less than 3 further on in the sequel.

STEP 2: Choose the arbitrary variables. This choice must be made in such a way that the rank of the matrix

147

formed by the coefficients of the dependent variables is 3. (This ensures that the A matrix is invertible. Note that in the above example, e <u>must</u> be chosen as an arbitrary variable to avoid a column of 0's in the A matrix.)

       STEP 3: <u>Solve the system of equations</u>. The solution is obtained by multiplying the augmented matrix by the inverse of the A matrix. We determine A^{-1} and perform the required multiplication using Minitab.

       STEP 4: <u>Assign convenient values to the arbitrary variables and determine the independent solutions</u>. In succession, assign each arbitrary variable the value 1 with all other arbitrary variables assigned the value 0. This process will yield as many independent solutions as there are arbitrary variables.

An Example

    Let's illustrate the procedure with a specific example involving a larger system than before:

$$0x_1 + 1x_2 + 2x_3 + 1x_4 + 1x_5 + 0x_6 + 0x_7 = 0$$

$$3x_1 + 0x_2 + 2x_3 - 3x_4 - 3x_5 + 1x_6 + 1x_7 = 0$$

$$0x_1 + 0x_2 - 2x_3 + 0x_4 + 0x_5 - 2x_6 + 0x_7 = 0$$

The augmented matrix is

$$
\begin{array}{ccccccc}
x_1 & x_2 & x_3 & x_4 & x_5 & x_6 & x_7 \\
\end{array}
$$
$$
\left[
\begin{array}{ccccccc|c}
0 & 1 & 2 & 1 & 1 & 0 & 0 & 0 \\
3 & 0 & 2 & -3 & -3 & 1 & 1 & 0 \\
0 & 0 & -2 & 0 & 0 & -2 & 0 & 0 \\
\end{array}
\right]
$$

STEP 1: The rank of the above system is 3 which is easily verified by showing that the determinant of the submatrix formed by the first three columns is nonzero. Thus, there are 7-3=4 arbitrary variables in the system.

STEP 2: Let's choose x_1, x_5, x_6 and x_7 as the arbitrary variables. Note in this example that both x_4 and x_5 cannot be dependent variables simultaneously since their columns are identical. Transposing the chosen arbitrary variables to the right-hand side yields the augmented matrix:

$$\begin{array}{ccccccc} x_2 & x_3 & x_4 & x_1 & x_5 & x_6 & x_7 \end{array}$$

$$\left[\begin{array}{ccc|cccc} 1 & 2 & 1 & 0 & -1 & 0 & 0 \\ 0 & 2 & -3 & -3 & 3 & -1 & -1 \\ 0 & -2 & 0 & 0 & 0 & 2 & 0 \end{array}\right]$$

$$\text{MATRIX A} \qquad\qquad \text{MATRIX B}$$

STEP 3: We solve the above system by multiplying the augmented matrix by the inverse of the A matrix.

$$\begin{array}{ccccccc} & & & x_2 & x_3 & x_4 & x_1 & x_5 & x_6 & x_7 \end{array}$$

$$\left[\begin{array}{ccc} 1 & 1/3 & 1/3 \\ 0 & 0 & -1/2 \\ 0 & -1/3 & -1/3 \end{array}\right] \left[\begin{array}{ccc|cccc} 1 & 0 & 1 & 0 & -1 & 0 & 0 \\ 0 & 2 & -3 & -3 & 3 & -1 & -1 \\ 0 & -2 & 0 & 0 & 0 & 2 & 0 \end{array}\right]$$

which yields,

$$\begin{array}{ccccccc} x_2 & x_3 & x_4 & x_1 & x_5 & x_6 & x_7 \end{array}$$

$$\left[\begin{array}{ccc|cccc} 1 & 0 & 0 & -1 & 0 & 1/3 & -1/3 \\ 0 & 1 & 0 & 0 & 0 & -1 & 0 \\ 0 & 0 & 1 & 1 & -1 & -1/3 & 1/3 \end{array}\right]$$

Assigning one of the arbitrary variables the value 1 with all other arbitrary variables the value 0 in succession yields the following four solutions:

$x_1 = 1$	$x_1 = 0$	$x_1 = 0$	$x_1 = 0$
$x_2 = -1$	$x_2 = 0$	$x_2 = 1/3$	$x_2 = -1/3$
$x_3 = 0$	$x_3 = 0$	$x_3 = -1$	$x_3 = 0$
$x_4 = 1$	$x_4 = -1$	$x_4 = -1/3$	$x_4 = 1/3$
$x_5 = 0$	$x_5 = 1$	$x_5 = 0$	$x_5 = 0$
$x_6 = 0$	$x_6 = 0$	$x_6 = 1$	$x_6 = 0$
$x_7 = 0$	$x_7 = 0$	$x_7 = 0$	$x_7 = 1$

Note that the values of the dependent variables are readily available as the last four columns of the final augmented matrix, as shown above.

The steps just illustrated are easily implemented using Minitab. We begin by discussing the Minitab commands required to enter a matrix into the system, copy a selected list of columns as a matrix, invert a matrix, and multiply matrices.

6.2   Matrix Entry And Matrix Operations

In Chapter 5 we investigated the use of the READ command to enter data into the rows of a matrix. In solving matrix equations of the form AX = B, the direct entering of the matrix A into the Minitab worksheet has its advantages. We begin by reviewing the READ command.

READ n by **n** matrix into **Mj**

For instance, the commands

```
MTB>READ 2 by 2 matrix into M1
DATA>1 2
DATA>3 4
MTB>PRINt M1
```

result in the following output:

MATRIX M1

1   2

3   4

In many applications the data to be used as entries in the desired matrix already exist as columns somewhere in your worksheet. A matrix can be formed easily from existing columns using the following COPY command:

MTB>COPY columns of the worksheet **Cj,...,Cj** into matrix **Mj**

For example, let's assume column C1 contains 1 2 3, C2 contains 1 -1 2, and C3 contains 0 -2 1. The Minitab commands

```
MTB>COPY C1 C3 C2 M2
MTB>PRIN M2
```

result in the following output:

MATRIX M2

1   0   1

2   -2   -1

3   1   2

The copy command is extremely useful because you can specify the order of the columns you have selected in forming your matrix. For example, to enter a new column C4 which will replace C2 in the third column above, use the following commands:

```
MTB>SET C4
DATA>-1 -1 -2
DATA>END
MTB>COPY C1 C3 C4 M3
MTB>PRIN M3
```

The resulting output will be the desired matrix:

MATRIX M3

1	0	-1
2	-2	-1
3	1	-2

The COPY command may also be used to copy the columns of an existing matrix into individual columns of your Minitab worksheet, or into another matrix:

```
MTB>COPY M1 into C1 C2
MTB>PRIN C1 C2
```

Here the output will be

ROW	C1	C3
1	1	2
2	3	4

We illustrate other uses of these commands in Exhibit 6.2.1 and apply them to executing a dimensional analysis in Section 6.3.

An Example Illustrating Minitab Matrix Entry

```
MTB > READ C1 C2 C3 C4 C5
DATA> 0 1 0 0 0
DATA> 0 0 1 1 0
DATA> 1 0 -2 0 0
DATA> END
      3 ROWS READ
MTB > PRINT C1-C5
 ROW    C1    C2    C3    C4    C5

   1     0     1     0     0     0
   2     0     0     1     1     0
   3     1     0    -2     0     0

MTB > COPY C1 C2 C3 C4 C5 INTO MATRIX M1
MTB > PRINT M1
MATRIX M1

   0    1    0    0    0
   0    0    1    1    0
   1    0   -2    0    0


MTB > COPY C2 C3 C4 INTO MATRIX M2
MTB > PRINT M2
MATRIX M2

   1    0    0
   0    1    1
   0   -2    0


MTB > SAVE 'PENDULUM'
```

Matrix Operations With Minitab

We now cover the Minitab matrix arithmetic operations of multiplication, addition, and subtraction, as well as the operations for finding the inverse and transpose of a matrix. The format for the Minitab commands INVERT, TRANSPOSE, MULTIPLY, ADD, and SUBTRACT is as follows:

MTB>CHOSEN MATRIX OPERATION Mj (to, from or by Mj Cj or Kj) put into Mj

We present five examples to cover each of the operations.

Example 1. Invert the matrix M1 and store it as M4:

MTB>**INVErt M1** put into **M4**

Example 2. Transpose the matrix M2 and store it as M5:

MTB>**TRANspose M2** put into **M5**

Example 3. Multiply matrix M3 times a matrix M4, a column C2, a stored constant K2, or a specified constant -1, and store the results as M6, M7, M8, and M9, respectively:

MTB>**MULTiply M3** by **M4** put into **M6**
MTB>**MULT M3 C2 M7**
MTB>**MULT M3 K2 M8**
MTB>**MULT M3 -1 M9**

Matrix multiplication is always performed in the order given. So in the above example, M6=M3*M4. The matrices must be conformable for the multiply operation.

Example 4. Add the matrices M7 and M8 and store the result as M9:

MTB>**ADD M7** to **M8** put into **M9**

Example 5. Subtract the matrix M2 from M3 and store the result as M6:

MTB>**SUBTract M2** from **M3** put into **M6**

Note that the order in which the matrices appear is important for the subtraction operation: we calculated M3 - M2 as opposed to M2 - M3.

6.2 EXERCISES

1. SET or READ the following data into your workspace:

a.
C1	C2	C3	C4	C5
0	1	0	0	0
0	0	1	1	0
1	0	-2	0	0

b.
C6	C7	C8	C9	C10	C11	C12
0	1	0	1	1	0	0
3	0	2	-3	-3	1	1
0	0	-2	0	0	-2	0

2. COPY the following columns in a matrix:

    a. C1 C2 C3 as matrix M1
    b. C2 C3 C4 as matrix M2
    c. C1 C2 C3 C4 C5 as matrix M3
    d. C7 C8 C9 as matrix M4

3. INVERT each matrix in Problem 2a, b, d.

4. MULTIPLY to find the following products, and print the result.

    a. M1*M2
    b. M2*M1
    c. M3*M2
    d. M4*C10

5. ADD to find the following sums, and print the result.

    a. M1+M2
    b. M2+M4
    c. M1+C4 (what happened?)

6. SUBTRACT to find the following differences, and print the result.

    a. M2-M1
    b. M4-M2
    c. M3-M1 (What happened?)

7. TRANSPOSE the matrices M2 and M3, and print the results.

8. Find the INVERSE of the matrices consisting of the designated columns, and print the results:

    a. C2 C3 C1
    b. C1 C3 C4
    c. C6 C8 C10
    d. C2 C8 C12

6.3   The Use Of Minitab In Dimensional Analysis And Similitude

The application of the process of dimensional analysis generally yields a system of equations having more unknowns than equations. Matrix operations can be used conveniently to find a complete set of dimensionless products $\{\Pi_1, \Pi_2, \ldots, \Pi_n\}$ as required by Buckingham's Theorem. We illustrate the procedure with the simple pendulum problem discussed in the introduction to this chapter.

As stated previously, we desire to approximate the relationship
$$t = f(m,g,r,\theta)$$
where the dimensions of the variables are given by the following table:

Variable	t	m	g	r	θ
Dimension	T	M	LT^{-2}	L	$M^0L^0T^0$

A product of the variables must be of the form

$$t^a m^b g^c r^d \theta^e$$

and forcing the product to be dimensionless requires

$$T^a M^b (LT^{-2})^c L^d (M^0 L^0 T^0)^e = M^0 L^0 T^0$$

or

$$M^b L^{c+d} T^{a-2c} = M^0 L^0 T^0.$$

Therefore, the product is dimensionless if and only if the exponents satisfy the following homogeneous system of equations:

$$M: \quad 0a + 1b + 0c + 0d + 0e = 0$$

$$L: \quad 0a + 0b + 1c + 1d + 0e = 0$$

$$T: \quad 1a + 0b - 2c + 0d + 0e = 0$$

This system can be written in matrix form as

$$\begin{bmatrix} 0 & 1 & 0 & 0 & 0 \\ 0 & 0 & 1 & 1 & 0 \\ 1 & 0 & -2 & 0 & 0 \end{bmatrix} \begin{bmatrix} a \\ b \\ c \\ d \\ e \end{bmatrix} = \begin{bmatrix} 0 \\ 0 \\ 0 \end{bmatrix}$$

As an illustration, we now implement the 4-step procedure of Section 6.1 using Minitab.

   STEP 1: <u>Determine the number of arbitrary variables</u>. The above system of equations has 5 unknowns and rank 3 yielding 5-3=2 arbitrary variables.

   STEP 2: <u>Choose the arbitrary variables</u>. Note that the column corresponding to the unknown e consists of all 0's; thus e <u>must</u> be chosen as an arbitrary variable. We choose the exponent a, corresponding to the dependent variable t, as the second arbitrary variable. (Other choices can be made, but they are not as convenient for carrying out the dimensional analysis process.)

155

STEP 3: <u>Solve the system of equations</u>. Assume that we have entered the coefficients of variables a through e in columns C1 through C5, respectively, using the READ or SET commands. Since we have chosen a and e as arbitrary variables, we first multiply those columns by -1 in preparation for transposing them to the right-hand side:

MTB>**MULT C1 -1 C1**
MTB>**MULT C5 -1 C5**

We then form our AUGMENTED MATRIX M1 with the columns corresponding to -a and -e transposed to the right-hand side:

MTB>**COPY C2 C3 C4 C1 C5 into matrix M1**

Next we need to invert the submatrix corresponding to the dependent variables b, c and d. Here this submatrix consists of the columns C2, C3 and C4.

MTB>**COPY C2 C3 C4 M2**
MTB>**INVErt M2 store as M3**

Finally we premultiply the entire augmented matrix by the inverse just found:

MTB>**MULTiply M3 by M1 store as M4**

The matrix M4 now contains the solution to the system. We print the result:

MTB>**PRINt M4**

For this example we get the augmented matrix:

$$
\begin{array}{ccccc}
b & c & d & a & e \\
\end{array}
$$
$$
\begin{bmatrix}
1 & 0 & 0 & 0 & 0 \\
0 & 1 & 0 & 0.5 & 0 \\
0 & 0 & 1 & -0.5 & 0
\end{bmatrix}
$$

If we have done our work correctly, the columns corresponding to the first three columns should be the identity matrix.

STEP 4: <u>Assign convenient values to the arbitrary variables and determine the solution</u>. This step can be achieved conveniently using the final augmented matrix M4: assigning a=1, e=0 yields b=0, c=0.5, d=-0.5 which corresponds to the product $t(g/r)^{1/2}$. The assignment a=0, e=1 yields the solution b = c = d = 0 which corresponds to a second dimensionless product θ. Applying Buckingham's Theorem then yields

$$H(t(g/r)^{1/2}, \theta) = 0$$

or, solving for t,

$$t = (r/g)^{1/2} h(\theta)$$

in agreement with our previous result.

Note that we could have used column headers merely by naming the appropriate columns and printing the columns instead of the matrix. This idea is demonstrated in Exhibit 6.3.1.

──────────────── EXHIBIT 6.3.1 ────────────────

Dimensional Analysis Of The Undamped Pendulum

```
MTB > RETR 'PENDULUM'
MTB > MULT C1 -1 C1
MTB > MULT C5 -1 C5
MTB > COPY C2 C3 C4 C1 C5 INTO M1
MTB > COPY C2 C3 C4    INTO M2
MTB > INVERT M2 M3
MTB > MULT M3 M1 M4
MTB > COPY M4 C2 C3 C4 C1 C5
MTB > NAME C1 'a',C2 'b',C3 'c',C4 'd',C5 'e'
MTB > PRINT C2 C3 C4 C1 C5
  ROW     b     c     d      a     e

    1     1     0     0    0.0     0
    2     0     1     0    0.5     0
    3     0     0     1   -0.5     0

MTB >
```

──

6.3   EXERCISES

1. The lift force F on a missile depends upon its length r, velocity v, diameter d, initial angle θ, density ρ, viscosity μ, gravity g, and speed of sound in air s. The dimension of each variable is

Variable	F	r	v	d	θ	ρ	μ	g	s
Dimension	MLT^{-2}	L	LT^{-1}	L	$M^0L^0T^0$	ML^{-3}	$ML^{-1}T^{-1}$	LT^{-2}	LT^{-1}

157

Use Minitab to show that

$$F = \rho v^2 r^2 \; h(d/r, \theta, \mu/\rho\, vr, s/v, rg/v^2)$$

where h is an arbitrary function to be approximated empirically. HINT: Choose F,d,θ,μ,s and g as arbitrary variables.

2. It is desired to study the velocity f of a fluid flowing in a smooth open channel. The velocity v is a function of length r, density ρ, viscosity μ, surface tension δ, and gravity g. Use Minitab to show that

$$v^2 = gr \; h(\rho vr/\mu, \rho v^2 r/\delta)$$

where h is an arbitrary function to be approximated empirically. HINT: Choose g, μ, and δ as the arbitrary variables. The dimension of surface tension is MT^{-2}.

6.4 Examples Illustrating Dimensional Analysis

Example 1 Explosion Analysis

Problem Identification: Predict the crater volume V produced by a spherical explosive located at some depth d in a particular soil medium.

Assumptions and Model Formulation: In order to characterize an explosive in more detail, 3 independent variables are needed: size, energy yield, and explosive density δ. The size can be given as charge mass W, charge energy E, or as the radius α of the spherical explosive. The energy yield can be either the specific energy Q_e or the energy density per unit volume Q_v. The following equations relate these variables:

$$W = E/Q_e$$

$$Q_v = \delta Q_e$$

$$\alpha^3 = 3W/4\pi\delta$$

One choice of these variables leads to the model formulation:

$$V = f(W, Q_e, \delta, \rho, g, d)$$

We desire to find all the dimensionless products of the form

$$v^a W^b Q_e{}^c \delta^e \rho^f g^k d^m$$

In Exhibit 6.4.1 we display a solution using Minitab. Note that:

1. The rank of the matrix is 3, so there are 7-3 = 4 arbitrary variables.

2. We have chosen v, ρ, g, d as the arbitrary variables. Note that we can interchange ρ and δ since they have the same dimension. Thus a set of dimensionless products is:

$$\Pi_1 = V\rho/W$$

$$\Pi_2 = (g/Q_e)(W/\delta)^{1/3}$$

$$\Pi_3 = d(\rho/W)^{1/3}$$

$$\Pi_4 = \rho/\delta$$

Application of Buckingham's Theorem yields $\Pi_1 = H(\Pi_2, \Pi_3, \Pi_4)$ or

$$V = (W/\rho)\ H((g/Q_e)(W/\delta)^{1/3},\ d(\rho/W)^{1/3},\ \rho/\delta)$$

───────────────── EXHIBIT 6.4.1 ─────────────────

Dimensional Analysis To Predict Crater Volume

```
MTB > READ C1-C7
DATA> 0  1  0  1  1  0  0
DATA> 3  0  2 -3 -3  1  1
DATA> 0  0 -2  0  0 -2  0
DATA> END
      3 ROWS READ
MTB > NAME  C1 'a',C2 'b',C3 'c',C4 'e',C5 'f',C6 'k',C7 'm'
MTB > PRINT C1-C7
 ROW    a    b    c    e    f    k    m

  1     0    1    0    1    1    0    0
  2     3    0    2   -3   -3    1    1
  3     0    0   -2    0    0   -2    0

MTB > MULT C1 -1 C1
MTB > MULT C5 -1 C5
MTB > MULT C6 -1 C6
MTB > MULT C7 -1 C7
MTB > COPY C2 C3 C4 C1 C5 C6 C7 INTO M1
MTB > COPY C2 C3 C4 INTO M2
MTB > INVERT M2 M3
MTB > MULT M3 M1 M4
MTB > COPY M4 INTO C2 C3 C4 C1 C5 C6 C7
MTB > PRINT C2 C3 C4 C1 C5 C6 C7
 ROW    b    c    e    a    f        k          m

  1     1    0    0   -1   -0   0.33333  -0.333333
  2     0    1    0    0    0  -1.00000   0.000000
  3     0    0    1    1   -1  -0.33333   0.333333

MTB > SAVE 'BOOM'
```

Example 2   How Long Should You Roast A Turkey?

One rule of thumb for roasting a turkey is the following: set the oven to 400 degrees F and allow 20 minutes per pound for cooking. How good is this rule?

Since most turkeys weigh between 10 and 20 pounds, we investigate the rule of thumb for those weights in Exhibit 6.4.2. Let's assume that the cooking time t is a function of the following variables:

ΔT_m: The difference between the temperatures of the raw meat and the oven, with dimension $ML^{-1}T^{-2}$

ΔT_c: The difference between the temperatures of the cooked meat and the oven, with dimension $ML^{-1}T^{-2}$

K: Coefficient of heat conduction, with dimension L^2T^{-1}

ℓ: A characteristic dimension of the turkey, assuming that turkeys are geometrically similar, with dimension L

We desire to find all dimensionless products of the form

$$t^a(\Delta T_m)^b(\Delta T_c)^c K^d \ell^e$$

This task is accomplished using Minitab in Exhibit 6.4.2. Note that the rank of the matrix is 3, requiring 5-3 = 2 arbitrary variables. We have chosen the dependent variable t and the variable ΔT_c as arbitrary, yielding the dimensionless products $\Pi_1 = tK/\ell^2$ and $\Pi_2 = \Delta T_c/\Delta T_m$. Application of Buckingham's Theorem then yields

$$t = (\ell^2/K)\ H(\Delta T_c/\Delta T_m)$$

Let V denote the volume of a turkey and W its weight. From our assumption that the turkeys are geometrically similar, we have $V \propto \ell^3$. If we further assume a constant weight density, then $W \propto \ell^3$, or $\ell^2 \propto W^{2/3}$. Now assume that K is constant for turkeys. Furthermore, if the ovens are at constant temperature and the raw meat is at room temperature (constant), then the ratio $\Delta T_c/\Delta T_m$ is constant. Under these assumptions

$$t \propto W^{2/3}$$

In Exhibit 6.4.2 we have plotted the rule of thumb against $W^{2/3}$ suggested from our dimensional analysis. The resulting plot reveals that the rule of thumb has no foundation. We should expect a proportionality rule of $t \propto W^{2/3}$.

EXHIBIT 6.4.2
Dimensional Analysis For Roasting A Turkey

```
MTB > SET C1
DATA> 10:20
DATA> END
MTB > LET C2=20*C1
MTB > NAME C1 'WEIGHT',C2 'TIME'
MTB > PLOT C2  0 450 C1 0 20
```

```
         TIME
        450.+
            -
            -
            -                                              * *
            -                                          *
            -                                    *   *
        300.+                                *
            -                         *   *
            -                   *   *
            -             *  *
            -        *   *
        150.+
            -
            -
            -
            -
          0.+
            +---------+---------+---------+---------+---------+WEIGHT
           0.0       4.0       8.0      12.0      16.0      20.0
```

```
MTB > READ C1-C5
DATA> 1  1  0  0  0
DATA> -1 -1  2  1  0
DATA> -2 -2 -1  0  1
DATA> END
     3 ROWS READ
MTB > NAME C1 'a',C2 'b',C3 'c',C4 'd',C5 'e'
MTB > PRINT C1-C5
 ROW    a      b      c      d      e

  1     1      1      0      0      0
  2    -1     -1      2      1      0
  3    -2     -2     -1      0      1
```

```
MTB > MULT C2 -1 C2
MTB > MULT C5 -1 C5
MTB > COPY C1 C3 C4 C2 C5 INTO M1
MTB > COPY C1 C3 C4 INTO M2
MTB > INVERT M2 M3
MTB > MULT M3 M1 M4
MTB > COPY M4 C1 C3 C4 C2 C5
MTB > PRINT C1 C3 C4 C2 C5
 ROW     a      c      d      b      e

   1      1      0      0     -1      0
   2      0      1      0      0      1
   3      0      0      1      0     -2

MTB > SAVE 'TURKEY'


MTB > SET TURKEY WEIGHT IN C1
DATA> 5 10 15 20
DATA> END
MTB > SET HOUR IN C2
DATA> 2 3.4 4.5 5.4
DATA> END
MTB > LET C3=C1**(2/3)
MTB > NAME C1 'WGHT T',C2 'HOUR',C3 'WGHT2/3'
MTB > PRINT C1 C2 C3
 ROW   WGHT T    HOUR    WGHT2/3

   1       5      2.0    2.92402
   2      10      3.4    4.64159
   3      15      4.5    6.08220
   4      20      5.4    7.36806


MTB > PLOT C2 0 7 VS C3 0 8
```

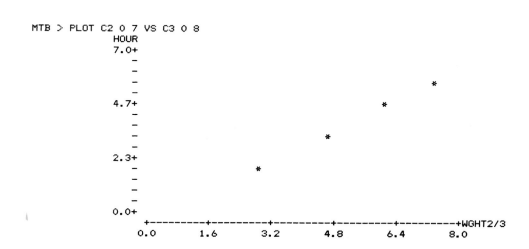

```
MTB > NOCONS
MTB > REGRESS C2 1 C3

THE REGRESSION EQUATION IS
HOUR = 0.732 W**2/3

                                    ST. DEV.      T-RATIO =
COLUMN        COEFFICIENT           OF COEF.      COEF/S.D.
NOCONSTANT
W**2/3        0.731505              0.007775        94.08

S = 0.08566

ANALYSIS OF VARIANCE

   DUE TO      DF             SS        MS=SS/DF
REGRESSION     1          64.948        64.948
RESIDUAL       3           0.022         0.007
TOTAL          4          64.970


DURBIN-WATSON STATISTIC = 1.11

MTB > SET C4
DATA> 1:20
DATA> END
MTB > LET C5=20*C4/60
MTB >


MTB > MPLOT C2 C1 C5 C4
          HOUR
          9.0+
             -
             -
             -
             -
             -                                            B B
          6.0+                                      B
             -                                 B B    B       A
             -                             B B
             -                         B B   A
             -                     2 B
          3.0+               B
             -           B B
             -        2 B
             -     B B
             -  B B
          0.0+
             +---------+---------+---------+---------+---------+WGHT T
             0.0       5.0      10.0      15.0      20.0      25.0
```

Example 3   Drag Force On A Submarine

In this example we consider the drag force F on a submarine to be a function of fluid velocity v, a characteristic dimension of the submarine r, fluid density ρ, fluid viscosity μ, and the speed of sound in the fluid c. Symbolically,

$$F = f(v, r, \rho, \mu, c)$$

We use dimensional analysis to relate the drag force of the submarine to the remaining variables. We then illustrate the use of the dimensionless products to determine how to build a scaled model of the prototype submarine. We desire dimensionless products of the form

$$F^a v^b r^d \rho^f \mu^h c^l$$

where

Variable	F	v	r	ρ	μ	c
Dimension	MLT^{-2}	LT^{-1}	L	ML^{-3}	$ML^{-1}T^{-1}$	LT^{-1}

A solution using Minitab is displayed in Exhibit 6.4.3. Note that there are $6 - 3 = 3$ arbitrary variables, chosen as F, μ and c. This choice gives the dimensionless products:

$$\Pi_1 = F/\rho v^2 r^2$$

$$\Pi_2 = vr\rho/\mu$$

$$\Pi_3 = v/c$$

Application of Buckingham's Theorem yields the solution:

$$F = \rho v^2 r^2 \; H(vr\rho/\mu, \; v/c)$$

To ensure that the scaled model resembles the prototype, we apply the design conditions:

$$(vr\rho/\mu)_m = (vr\rho/\mu)_p$$

and

$$(v/c)_m = (v/c)_p$$

where the subscript m denotes the scaled model and p the prototype.

Dimensional Analysis For Drag Force On A Submarine

```
MTB > RETR 'SUB'
MTB > READ C1-C6
DATA> 1 0 0 1 1 0
DATA> 1 1 1 -3 -1 1
DATA> -2 -1 0 0 -1 -1
DATA> END
      3 ROWS READ
MTB > PRINT C1-C6
 ROW      a      b      d      f      h      l

  1       1      0      0      1      1      0
  2       1      1      1     -3     -1      1
  3      -2     -1      0      0     -1     -1

MTB > MULT C1 -1 C1
MTB > MULT C5 -1 C5
MTB > MULT C6 -1 C6
MTB > COPY C2 C3 C4 C1 C5 C6 INTO M1
MTB > COPY C2 C3 C4 INTO M2
MTB > INVERT M2 M3
MTB > MULT M3 M1 M4
MTB > COPY M4 INTO C2 C3 C4 C1 C5 C6
MTB > PRINT C2 C3 C4 C1 C5 C6
 ROW      b      d      f      a      h      l

  1       1      0      0     -2     -1     -1
  2       0      1      0     -2     -1      0
  3       0      0      1     -1     -1      0
```

6.4   PROJECTS

The following project refers to the explosion analysis example.

Given $V = f(W, Q_e, \delta, \rho, g, d)$ and the accompanying table:

Variable	V	W	Q_e	δ	ρ	g	d
Dimension	L^3	M	L^2T^{-2}	ML^{-3}	ML^{-3}	LT^{-2}	L

a)   Determine a complete set of dimensionless products and apply Buckingham's Theorem.   Choose V, ρ, g, and d to be the arbitrary variables.

b) Assume ρ is essentially constant for the soils being used, and restrict the explosive type to TNT. Under these conditions ρ/δ is essentially constant, yielding

$$\Pi_1 = f(\Pi_2, \Pi_3)$$

You have collected the following data with $\Pi_2 = 1.5 \times 10^{-6}$

Π_3	0	2	4	6	8	10	12	14
Π_1	15	150	425	750	825	425	250	90

i. Construct a scatter plot of Π_1 versus Π_3. Does a trend exist?

ii. How accurate do you think the data are? Find an empirical model that captures the trend of the data with the accuracy commensurate with your appraisal of the accuracy of the data.

iii. Use your empirical model to predict the volume of a crater using TNT in desert alluvium with (CGS system) W = 1500 g, ρ = 1.53 g/cm^3, and Π_3 = 12.5.

7

SIMULATION

```
RANDOM NUMBER GENERATION:
          URANDOM
          IRANDOM
          DRANDOM

NEGATIVITY:
          SIGNS
```

A modeler may encounter situations where the construction of an analytic model is infeasible due to the complexity of the situation. In Chapter 5 we learned how to fit empirical models to a set of observed data and in Chapter 6 we learned how to use dimensional analysis to reduce the number of experiments that must be conducted when behavior is duplicated experimentally. In instances where the behavior cannot be modeled analytically, or data collected directly, the modeler might simulate the behavior indirectly in some manner, and then test various alternatives being considered to estimate how each affects the behavior. Data can then be collected to determine which alternative is best. An example of one type of simulation is a scaled model of a jet aircraft in a wind tunnel used to estimate the effects of very high speeds for various designs of the aircraft. Another type of simulation is Monte Carlo simulation and is typically accomplished with the aid of a computer. In this chapter we learn how to use Minitab to construct Monte Carlo simulations of deterministic and probabilistic behavior. There are a number of serious mathematical concerns associated with the construction and interpretation of Monte Carlo simulations (see Giordano and Weir, Chapter 8, op. cit.); here we discuss only the Minitab implementation of a Monte Carlo simulation.

7.1 Deterministic Behavior

In this section we illustrate the use of Monte Carlo simulation to model a deterministic behavior, the area under a nonnegative curve. We begin by finding an approximate value to the area under a nonnegative curve. Specifically, suppose $y = f(x)$ is some given continuous function satisfying $0 \leq f(x) < M$ over the closed

interval a ≤ x ≤ b.   Here the number M is simply some positive
constant that bounds the function.   This situation is depicted in
Figure 7-1.   Notice that the area we seek is totally contained
within the rectangular region of height M and width b-a (the length
of the interval over which f is defined).

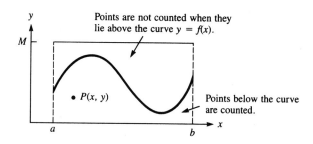

FIGURE 7-1   The area under the nonnegative curve y = f(x) is
contained within the rectangle of height M and base length b-a.

Now generate a point P(x,y) at random from within the rectangular region.   We do this by generating a random number for x
satisfying a ≤ x ≤ b and a random number for y satisfying
0 ≤ y ≤ M.   Now we ask if the random point P(x,y) so generated
lies on or below the curve.   That is, does the y-coordinate satisfy
0 ≤ y ≤ f(x)?   If the answer is "yes," then we count the point P.
Next we generate a new random point in the same way and repeat the
question.   Continuing to generate random points in this fashion an
approximate value for the area under the curve can be computed
using the following formula:

$$\frac{\text{area under curve}}{\text{area of rectangle}} \approx \frac{\text{number of points counted below the curve}}{\text{total number of random points}}$$

We need to generate a fairly large number of trials to obtain a
reasonable approximation for the area under the curve.   However,
because the method is strictly probabilistic, you will not necessarily achieve greater numerical accuracy using a run of 10000
trials instead of a run of 1000 trials because the particular run
of random numbers may not be that good.   The following algorithm
gives the sequence of calculations needed for a general computer
simulation of this Monte Carlo technique to find the area under a
curve.   We next present the necessary Minitab commands before
constructing the simulation.

MONTE CARLO ALGORITHM
TO CALCULATE AREA UNDER AN ARBITRARY CURVE

Input	Total number n of random points to be generated. Endpoints a and b for the interval $a \leq x \leq b$. Upper bound M for the positive function $y = f(x)$.
Output	Monte Carlo approximation to the area under the curve $y = f(x)$ over the interval $a \leq x \leq b$.
Step 1	Specify the function f.
Step 2	Calculate n random coordinates x_i and y_i where $a \leq x_i \leq b$ and $0 \leq y \leq M$.
Step 3	Calculate all values $f(x_i)$ for all n random coordinates.
Step 4	Calculate all differences $y_i - f(x_i)$.
Step 5	COUNT the number of 0's and negative values among the differences $y_i - f(x_i)$.
Step 6	Calculate the AREA = M(b - a)*COUNT/n.
Step 7	Output AREA.

The first Minitab command we introduce is the command URANDOM whose format is:

MTB>**URAN**dom **K** observations, put into column **Cj**

The above command will generate K random numbers from the interval (0,1) and place them in column Cj. These random numbers can then be scaled to the desired values. For example, suppose you want to generate 10 values for x from the interval (-4,4). The first command would be:

MTB>**URAN**dom **10 C1**

You now have 10 values between 0 and 1 stored in column C1. Next these values must be scaled to the desired interval (-4,4). In general, to scale x values in the interval (0,1) to the interval (a,b), use the transformation $z = b - (b-a)x$. Thus, the equation $z = 4 - 8x$ scales the values in the interval (0,1) to the interval (-4,4). The appropriate Minitab command to complete the scaling would be:

MTB>**LET C2 = 4 - (8 * C1)**

The scaled values for x now reside in column C2.

An Example: The Area Under The Cosine Curve

We are now prepared to generate pairs of random numbers to approximate the area under a nonnegative curve. The curve we use is the cosine curve for values of x between $-\pi/2$ and $\pi/2$. As an upper bound for the value of cos x we take the number 2; that is, we know $0 \leq \cos x \leq 2$.   The following series of Minitab commands create the desired random numbers:

```
MTB>LET K1=100 (desired quantity of random numbers)
MTB>URANdom K1,C1 (x values in column C1)
MTB>URANdom K1,C2 (y values in column C2)
MTB>LET C3=3.14159/2 - 3.14159*C1 (scaling x to its interval)
MTB>NAME C3 'Xi'
MTB>LET C4 = 2 - (2 - 0)*C2 (scaling y to its (0,2) interval)
MTB>NAME C4 'Yi'
```

We now have values for x in the interval $(-\pi/2, \pi/2)$ stored in column C3 and values for y in the interval (0,2) stored in column C4.   Next we need to create values for f(x).   We create the function f(x) = cos x with the following commands:

```
MTB>LET C5 = cos ('Xi')
MTB>NAME C5 'f(Xi)'
```

Since Minitab does not have the capability of conditional statements, we need to use a series of commands to test the (x,y) pairs to determine if they lie on or below the designated curve. The statement we need to test is:

$$Yi \leq f(Xi) \text{ or, equivalently, } Yi - f(Xi) \leq 0?$$

We compute Yi - f(Xi) using the LET command.

```
MTB>LET C6 = 'Yi' - 'F(Xi)'
MTB>NAME C6 'DIFF'
```

In the column "DIFF" we have negative, positive and zero values. The positive values represent (x,y) pairs that lie above the curve. If any 0 values are present they lie on the curve, and any negative values lie below the curve. The value of K1 represents the total number of points generated.   We need to count the number of points on or below the curve so we use the SIGNS command to check for positive, negative and zero values.   The format for this command is:

```
MTB>SIGNS Cj1 Cj2
```

The command checks column Cj1 and assigns a value of -1 for each negative number, 0 for each zero, and 1 for each positive number, placing these corresponding values in column Cj2.   We now

170

use this command to check for points that lie on or below the curve.

MTB>**SIGNS** 'DIFF' C7

All values in the 'DIFF' column have been identified as positive, negative, or zero with the results recorded in column C7. We count the number of 0's and -1's in C7 by eliminating the +1's from the column and then counting the number of remaining values (0's and 1's). The OMIT command eliminates all the +1's and the COUNT command completes the procedure:

MTB>**OMIT** 1 C7 C8
MTB>**COUNT** C8 K5

We now have counted the number of points on or below the curve and can use the following equation to approximate the area under the curve:

$$\text{AREA} = \frac{(\text{area of rectangle})(\text{number of points on or below curve})}{\text{total number of points generated}}$$

The area of the rectangle is 2π, the number of points on or below the curve is K5, and the total number of points generated is K1. Thus, the required command is:

MTB>**LET** K6 = 2 * 3.14159 * K5 / K1

The approximate area under the curve is the value of K6.

A macro for computing the area under a nonnegative curve is given in Exhibit 7.1. The macro is written in terms of the constants defined below:

    K1: the number of random points to generate
    K2: the right endpoint for the x interval
    K3: the left endpoint for the x interval
    K4: an upper bound for y
    K5: a lower bound for y
    K6: the number of points on or below the curve
    K7: the approximate area

In order to use the macro, the constants K1,K2,K3,K4 and K5 must be entered before execution of the program. The macro assumes that the function named 'FUNC' is stored as the macro.

In Exhibit 7.1 the macro 'AREA' is presented first. Then we store a function, enter constants, and finally execute the macro.

A Monte Carlo Method To Find The Area Under A
Nonnegative Curve y = f(x), a ≤ x ≤ b

```
MTB > STORE 'AREA'
STOR> NOPRINT
STOR> NOECHO
STOR> URAN K1,C1
STOR> LET C2=K3-((K3-K2)*C1)
STOR> URAN K1,C3
STOR> LET C4=K4-((K4-K5)*C3)
STOR> NAME C2 'XI',C4 'YI'
STOR> EXEC 'FUNC'
STOR> NAME C5 'F(XI)'
STOR> LET C6='YI'-'F(XI)'
STOR> NAME C6 'DIFF'
STOR> SIGNS 'DIFF',C7
STOR> OMIT 1 C7,C8
STOR> COUNT C8,K6
STOR> LET K7=(K4-K5)*(K3-K2)*K6/K1
STOR> PRINT K7
STOR> ECHO
STOR> END
MTB >


MTB > STORE 'FUNC'
STOR> NOECHO
STOR> LET C5=COS('XI')
STOR> END
MTB >


MTB > NOTE YOU MUST ENTER THE NUMBER OF TRIALS AND THE BOUNDS.
MTB > LET K1=1000
MTB > LET K2=-3.14159/2
MTB > LET K3=3.14159/2
MTB > LET K4=2.
MTB > LET K5=0.0
MTB > EXEC 'AREA'
MTB > NOPRINT
    343 NEGATIVE VALUES       0  ZERO VALUES    657  POSITIVE VALUES

    COUNT    =        343.00
K7         2.15513
MTB > END
MTB >
```

```
MTB > EXEC 'AREA' 5
MTB > NOPRINT
    305 NEGATIVE VALUES      0   ZERO VALUES    695  POSITIVE VALUES

    COUNT    =       305.00
K7       1.91637
MTB > END
MTB > NOPRINT
    315 NEGATIVE VALUES      0   ZERO VALUES    685  POSITIVE VALUES

    COUNT    =       315.00
K7       1.97920
MTB > END
MTB > NOPRINT
    312 NEGATIVE VALUES      0   ZERO VALUES    688  POSITIVE VALUES

    COUNT    =       312.00
K7       1.96035
MTB > END
MTB > NOPRINT
    318 NEGATIVE VALUES      0   ZERO VALUES    682  POSITIVE VALUES

    COUNT    =       318.00
K7       1.99805
MTB > END
MTB > NOPRINT
    302 NEGATIVE VALUES      0   ZERO VALUES    698  POSITIVE VALUES

    COUNT    =       302.00
K7       1.89752
MTB > END
MTB >


MTB > NOTE TO CHANGE THE NUMBER OF TRIALS CHANGE K1,LET K1= A NEW NUMBER.
MTB > NOTE TO CHANGE THE FUNCTION,CHANGE STORE FUNC.
MTB >


MTB > LET K1=10000
MTB > EXEC 'AREA'
MTB > NOPRINT
   3241 NEGATIVE VALUES      1   ZERO VALUES   6758  POSITIVE VALUES

    COUNT    =       3242.0
K7       2.03701
MTB > END
MTB >
```

```
MTB > EXEC 'AREA' 5
MTB > NOPRINT
   3138 NEGATIVE VALUES        0  ZERO VALUES    6862  POSITIVE VALUES

    COUNT    =      3138.0
K7       1.97166
MTB > END
MTB > NOPRINT
   3176 NEGATIVE VALUES        0  ZERO VALUES    6824  POSITIVE VALUES

    COUNT    =      3176.0
K7       1.99554
MTB > END
MTB > NOPRINT
   3211 NEGATIVE VALUES        1  ZERO VALUES    6788  POSITIVE VALUES

    COUNT    =      3212.0
K7       2.01816
MTB > END
MTB > NOPRINT
   3090 NEGATIVE VALUES        0  ZERO VALUES    6910  POSITIVE VALUES

    COUNT    =      3090.0
K7       1.94150
MTB > END
MTB > NOPRINT
   3236 NEGATIVE VALUES        1  ZERO VALUES    6763  POSITIVE VALUES

    COUNT    =      3237.0
K7       2.03386
MTB > END
```

7.1 EXERCISES

1. Generate 100 uniform random numbers in each of the following intervals:

    (a)   (0,1)
    (b)   (-1,1)
    (c)   (.5,2)
    (d)   (10,100)
    (e)   (1,6)

2. Use Monte Carlo simulation to approximate the area under the curve $y = 1 + \sin x$ over the interval $-\pi/2 < x < \pi/2$. Use the number 2 as an upper bound for y.

3. Use Monte Carlo Simulation to approximate π by considering the percentage of points generated randomly inside the unit quarter circle

$$Q: x^2 + y^2 = 1, \quad x \geq 0, \quad y \geq 0$$

when the quarter circle is taken to be inside the square

$$S: 0 \leq x \leq 1 \quad \text{and} \quad 0 \leq y \leq 1.$$

Use the approximation $\pi/4$ = area Q/area S.

4. Use Monte Carlo simulation to find the volume under the sphere $x^2 + y^2 + z^2 \leq 1$ that lies in the 1st octant.

7.2   Probabilistic Behavior

In Monte Carlo simulation a sequence of random numbers can be used to simulate a probabilistic experiment. We will illustrate a Monte Carlo simulation with a simple probabilistic process, the roll of a die. There are 2 possible situations: that of a fair die and the case of an unfair die.

Case 1   A Fair Die

There are 6 possible outcomes when you roll a fair die, all occurring with probability 1/6. For each side of the die we will generate a corresponding integer random number. In Minitab the following command is used to generate integer random numbers:

MTB>**IRAN**dom **K1** integers between **K2** and **K3** store in **Cj**

In this command, K1 represents the number of random integers to be generated, K2 and K3 represent the lower and upper limits of the random integers, respectively, and Cj denotes the column in which the random integers are stored. For example, the command required to simulate 100 rolls of a standard die would be:

MTB>**IRAN**dom **100 1 6 C1**

Minitab automatically prints the random numbers generated. To avoid this printing, type NOPRINT before using any random number generator.

The IRANdom command generates integers over the interval specified according to the uniform distribution. That is, each integer has an equal likelihood of occurring.

In Exhibit 7.2.1 we first illustrate the simulation of rolling a fair die 100 times, and then 10000 times. We count the number of 1's, 2's, ..., 6's using the CHOOSE and COUNT commands. Note that with 10000 rolls the percentage of occurrence of each side of the die is reasonably close to 1/6.

EXHIBIT 7.2.1
Monte Carlo Simulation For Rolling A Fair Die

```
MTB > STORE 'DICE'
STOR> IRANDOM K1 1 6 C1
STOR> CHOOSE 1 C1 C10
STOR> CHOOSE 2 C1 C11
STOR> CHOOSE 3 C1 C12
STOR> CHOOSE 4 C1 C13
STOR> CHOOSE 5 C1 C14
STOR> CHOOSE 6 C1 C15
STOR> COUNT C10
STOR> COUNT C11
STOR> COUNT C12
STOR> COUNT C13
STOR> COUNT C14
STOR> COUNT C15
STOR> END


MTB > LET K1=100
MTB > EXEC 'DICE'
MTB > IRANDOM K1 1 6 C1
MTB > CHOOSE 1 C1 C10
MTB > CHOOSE 2 C1 C11
MTB > CHOOSE 3 C1 C12
MTB > CHOOSE 4 C1 C13
MTB > CHOOSE 5 C1 C14
MTB > CHOOSE 6 C1 C15
MTB > COUNT C10
   COUNT    =       16.000
MTB > COUNT C11
   COUNT    =       12.000
MTB > COUNT C12
   COUNT    =       14.000
MTB > COUNT C13
   COUNT    =       17.000
MTB > COUNT C14
   COUNT    =       21.000
MTB > COUNT C15
   COUNT    =       20.000
MTB > END
```

```
MTB > EXEC 'DICE'
MTB > IRANDOM K1 1 6 C1
MTB > CHOOSE 1 C1 C10
MTB > CHOOSE 2 C1 C11
MTB > CHOOSE 3 C1 C12
MTB > CHOOSE 4 C1 C13
MTB > CHOOSE 5 C1 C14
MTB > CHOOSE 6 C1 C15
MTB > COUNT C10
    COUNT    =        13.000
MTB > COUNT C11
    COUNT    =        12.000
MTB > COUNT C12
    COUNT    =        21.000
MTB > COUNT C13
    COUNT    =        17.000
MTB > COUNT C14
    COUNT    =        20.000
MTB > COUNT C15
    COUNT    =        17.000
MTB > END
MTB >


MTB > LET K1=10000
MTB > EXEC 'DICE'
MTB > IRANDOM K1 1 6 C1
MTB > CHOOSE 1 C1 C10
MTB > CHOOSE 2 C1 C11
MTB > CHOOSE 3 C1 C12
MTB > CHOOSE 4 C1 C13
MTB > CHOOSE 5 C1 C14
MTB > CHOOSE 6 C1 C15
MTB > COUNT C10
    COUNT    =        1706.0
MTB > COUNT C11
    COUNT    =        1697.0
MTB > COUNT C12
    COUNT    =        1681.0
MTB > COUNT C13
    COUNT    =        1639.0
MTB > COUNT C14
    COUNT    =        1603.0
MTB > COUNT C15
    COUNT    =        1674.0
MTB > END
```

Case 2   An Unfair Die

Suppose someone has tampered with the die to make it unfair. That is, the probabilities of outcomes for each side are no longer equal. Assume that these probabilities are given as follows:

$$Probability(side\ 1) = P(1) = .10$$
$$Probability(side\ 2) = P(2) = .20$$
$$Probability(side\ 3) = P(3) = .15$$
$$Probability(side\ 4) = P(4) = .25$$
$$Probability(side\ 5) = P(5) = .20$$
$$Probability(side\ 6) = P(6) = .10$$

To generate random numbers according to a discrete distribution such as above, use the Minitab command DRANDOM with the following format:

MTB>**DRAN**dom **K1** observations, possible outcomes in **Cj1**,
     corresponding probabilities in **Cj2**, put into **Cj3**

To simulate the unfair die described above, first put the possible outcomes 1,2,3,4,5,6 in one column and the corresponding probabilities in a second column. The DRANDOM command then stores the outcomes for the specified number of K1 observations in a third column. We illustrate the simulation in Exhibit 7.2.2 where we also count the number of occurrences of each face using the CHOOSE and COUNT commands described earlier.

──────────── **EXHIBIT 7.2.2** ────────────

Monte Carlo Simulation Of Rolling An Unfair Die

```
MTB > SET C1
DATA> 1 2 3 4 5 6
DATA> END
MTB > SET C2
DATA> .1 .2 .15 .25 .2 .1
DATA> END
MTB > NAME C1 'OUTCOME',C2 'PROB'
MTB > PRINT C1 C2
 ROW  OUTCOME    PROB

   1        1    0.10
   2        2    0.20
   3        3    0.15
   4        4    0.25
   5        5    0.20
   6        6    0.10
```

178

```
MTB > STORE 'UNEQDIE'
STOR> DRAN K1,C1,C2,C3
STOR> CHOOSE 1 C3 C10
STOR> CHOOSE 2 C3 C11
STOR> CHOOSE 3 C3 C12
STOR> CHOOSE 4 C3 C13
STOR> CHOOSE 5 C3 C14
STOR> CHOOSE 6 C3 C15
STOR> COUNT C10
STOR> COUNT C11
STOR> COUNT C12
STOR> COUNT C13
STOR> COUNT C14
STOR> COUNT C15
STOR> END
MTB >


MTB > LET K1=100
MTB > EXEC 'UNEQDIE'
MTB > DRAN K1,C1,C2,C3
MTB > CHOOSE 1 C3 C10
MTB > CHOOSE 2 C3 C11
MTB > CHOOSE 3 C3 C12
MTB > CHOOSE 4 C3 C13
MTB > CHOOSE 5 C3 C14
MTB > CHOOSE 6 C3 C15
MTB > COUNT C10
   COUNT    =        7.0000
MTB > COUNT C11
   COUNT    =        16.000
MTB > COUNT C12
   COUNT    =        15.000
MTB > COUNT C13
   COUNT    =        21.000
MTB > COUNT C14
   COUNT    =        31.000
MTB > COUNT C15
   COUNT    =        10.000
MTB > END
MTB >
```

```
MTB > LET K1=10000
MTB > EXEC 'UNEQDIE'
MTB > DRAN K1,C1,C2,C3
MTB > CHOOSE 1 C3 C10
MTB > CHOOSE 2 C3 C11
MTB > CHOOSE 3 C3 C12
MTB > CHOOSE 4 C3 C13
MTB > CHOOSE 5 C3 C14
MTB > CHOOSE 6 C3 C15
MTB > COUNT C10
   COUNT    =        989.00
MTB > COUNT C11
   COUNT    =        1978.0
MTB > COUNT C12
   COUNT    =        1440.0
MTB > COUNT C13
   COUNT    =        2579.0
MTB > COUNT C14
   COUNT    =        1971.0
MTB > COUNT C15
   COUNT    =        1043.0
MTB > END
```

7.2 EXERCISES

1. Simulate 1000 trials of the flip of a fair coin.

2. Simulate 1000 trials of the roll of a fair die. How many occurrences of each side of a die do you expect to see?

3. Consider the unfair die described below. Simulate 1000 trials of rolling this die. Count the number of occurrences of each side.

$$P(1) = .20$$
$$P(2) = .05$$
$$P(3) = .35$$
$$P(4) = .10$$
$$P(5) = .15$$
$$P(6) = .15$$

4. Consider an unfair die with 10 sides described by the probabilities below. Simulate 1000 and 10000 rolls of this die and compare your results for the 2 trials.

$$P(1) = .05$$
$$P(2) = .03$$
$$P(3) = .12$$
$$P(4) = .10$$
$$P(5) = .05$$

```
P(6)  = .15
P(7)  = .20
P(8)  = .05
P(9)  = .15
P(10) = .10
```

7.3   An Illustrative Example: An Inventory Model

Background: You are a consultant to an owner of gasoline stations along a superhighway.   The owner wants to maximize his profits and meet the consumer demands at each station, so you have identified the following problem.

Problem: Minimize the average daily cost (ADC) of delivering and storing sufficient gasoline at each station to meet consumer demand.

Assumptions: For an initial model, you consider that the average daily cost is a function of demand rate, storage costs, and delivery costs.   Your analysis reveals that in the short run you can accept the delivery and storage costs as constants.   You desire to build a submodel for the demand rate and begin by collecting historical data for demand at the gasoline station being studied. The data collected for 1000 days is shown in Table 7.3-1.

TABLE 7.3-1

History Of Demand At A Particular Gasoline Station

Number of gallons demanded	Number of occurrences (in days)
1000-1099	10
1100-1199	20
1200-1299	50
1300-1399	120
1400-1499	200
1500-1599	270
1600-1699	180
1700-1799	80
1800-1899	40
1900-1999	30
	1000

Model Formulation: The probabilities of each demand level are calculated and displayed in Table 7.3-2.   As an initial approximation, we will accept the midpoints of each demand interval as the number of gallons demanded to simulate in this example.   The midpoints and the corresponding probabilities of occurrences are given in Table 7.3-3.

TABLE 7.3-2

Probability Of Occurrence Of Each Demand Level

Number of gallons demanded	Probability of occurrence
1000-1099	0.01
1100-1199	0.02
1200-1299	0.05
1300-1399	0.12
1400-1499	0.20
1500-1599	0.27
1600-1699	0.18
1700-1799	0.08
1800-1899	0.04
1900-1999	0.03
	1.00

TABLE 7.3-3

Probability Of Occurrence Of Each Demand Level
Given At The Midpoint Of That Level

Number of gallons demanded	Probability of occurrence
1050	0.01
1150	0.02
1250	0.05
1350	0.12
1450	0.20
1550	0.27
1650	0.18
1750	0.08
1850	0.04
1950	0.03
	1.00

A plot of the probability of occurrence versus daily demand is obtained by Minitab and displayed in Figure 7-2. As part of our simulation we wish to generate daily demands that follow the distribution observed in the historical data plotted in the figure. By simulating typical daily demand patterns, we can estimate the average daily cost likely to be incurred when following different inventory strategies. That is, we can estimate the effect of varying either the order quantity or the time between deliveries. The average daily cost of various strategies can be computed using the equations.

COST = delivery costs + storage costs

Average Daily Cost = COST/time in days

The computations necessary to obtain the average daily cost are shown in the Inventory Algorithm below.

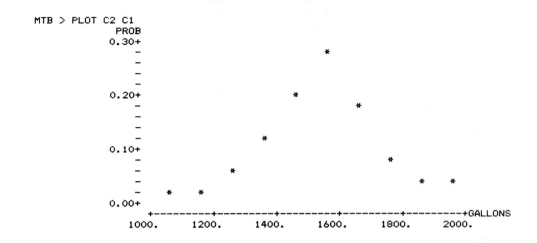

```
MTB > PLOT C2 C1
         PROB
        0.30+
             -                                           *
             -
             -
             -
        0.20+                                 *
             -                                      *
             -
             -
             -                        *
        0.10+
             -                                                *
             -                *
             -                                                   *    *
             -        *    *
        0.00+
             +---------+---------+---------+---------+---------+GALLONS
           1000.     1200.     1400.     1600.     1800.     2000.
```

FIGURE 7-2 Minitab plot of the probability of occurrence versus the demand level for the inventory example

Summary of Monte Carlo Inventory Algorithm Terms

Q: delivery quantity of gasoline in gallons
T: time between deliveries in days
I: current inventory in gallons
d: delivery cost in dollars per delivery
s: storage cost per gallon per day
C: total running cost
c: average daily cost
N: number of days to run the simulation
q_i: a daily demand

MONTE CARLO INVENTORY ALGORITHM

<u>Input</u>	Q,T,d,s,N
<u>Output</u>	c
<u>Step 1</u>	Initialize:
	$I = 0$
	$C = 0$

Step 2	Begin the next inventory cycle with a delivery:

$$I = I + Q$$
$$C = C + d$$

Step 3	Simulate each day in the inventory cycle:

For $i = 1,2,\ldots,T$, do Steps 4-6

Step 4	Generate a demand q_i using DRANDOM command.
Step 5	Update the current inventory: $I = I - q_i$.
Step 6	Compute the daily storage cost and total running cost, unless the inventory has been depleted: If $I \leq 0$, then set $I = 0$ and GOTO Step 7. Else $C = C + I * s$.
Step 7	Return to Step 2 until the desired number of cycles, which equals the integer portion of (N/T), are simulated for the current strategy.
Step 8	Compute the average daily cost: $c = C/N$.
Step 9	Output c.
	STOP

In the Minitab implementation of the above algorithm we use the following constants:

K1	=	Q, the delivery quantity
K2	=	T, the time between deliveries
K3	=	d, the delivery cost
K4	=	s, the storage cost
K5	=	N, the number of days to simulate
K6	=	I, the current inventory
K7	=	C, the total running cost
K8	=	C/N, the average daily cost
K10	=	Int (N/T), the number of cycles in the simulation

The Minitab implementation consists of several macros which are now explained. It is assumed that the observed demands are stored in column C1 and their corresponding probability of occurrence in column C2.

Step 2, a delivery, is coded as follows:

```
MTB>STOR 'DELIVER'
MTB>LET K6 = K6 + K1
MTB>LET K7 = K7 + K3
```

Steps 4-6, a day in the inventory cycle:

```
MTB<STOR 'CYCLE'
MTB>NOPRint
MTB>DRAN 1 C1 C2 C10
MTB>LET K6 = K6 - C10
MTB>LET K7 = K7 + (K6 * K4)
MTB>END
```

Step 3, an entire inventory cycle with the average daily cost computed at the end of the cycle:

```
MTB>STOR 'INVENTORY'
MTB>EXEC 'DELIVER'
MTB>EXEC 'CYCLE' K2 times
MTB>LET K8 = K7/K5
MTB>END
```

Step 7, repetition of the desired number of inventory cycles, and Step 9, output of the average daily cost and final inventory level:

```
MTB>STOR 'MINCOST'
MTB>NOECho
MTB>EXEC 'INVENTORY' K10 times
MTB>PRINt K8,K6
MTB>LET K6 = 0
MTB>LET K7 = 0
MTB>ECHO
MTB>END
```

Execution of the program is demonstrated in Exhibit 7.3.1.

Sensitivity Analysis Of The Inventory Simulation

A sensitivity analysis can be accomplished to approximate the best combination of Q and T in order to minimize the average daily cost while still meeting demand. For example, in Exhibit 7.3.1 we used various values of Q. In the initial run the following parameter values were used: Q = 11500, T = 7, d = 500, and s = 0.05. Allowing Q to vary, we then attempt to find a value of Q that meets demand at approximately the minimal average daily cost. If the gasoline station owner finds that he or she is better off ordering a certain amount Q of gasoline because of a contract with the distributor or a volume discount, it may be desirable instead to optimize T, the time between deliveries. This optimization can be achieved following a similar procedure to that displayed in Exhibit 7.3.1.

EXHIBIT 7.3.1

Program To Simulate Minimizing The Average Daily Cost
Of Delivering And Storing Gasoline

```
MTB > READ C1 C2
DATA> 1050 .01
DATA> 1150 .02
DATA> 1250 .05
DATA> 1350 .12
DATA> 1450 .20
DATA> 1550 .27
DATA> 1650 .18
DATA> 1750 .08
DATA> 1850 .04
DATA> 1950 .03
DATA> END
     10 ROWS READ
MTB > NAME C1 'DEMAND',C2 'PROB'


PRINT C1 C2 K1-K7,K10
  K1         11500.0
  K2         7.00000
  K3         500.000
  K4         0.0500000
  K5         35.0000
  K6         0
  K7         0
  K10        5.00000
  ROW    DEMAND    PROB

    1      1050     0.01
    2      1150     0.02
    3      1250     0.05
    4      1350     0.12
    5      1450     0.20
    6      1550     0.27
    7      1650     0.18
    8      1750     0.08
    9      1850     0.04
   10      1950     0.03
```

```
MTB > EXEC 'MINCOST' 5
K8         384.427
K6         2850.00
MTB > END
K8         463.999
K6         5950.00
MTB > END
K8         400.284
K6         4150.00
MTB > END
K8         441.999
K6         3750.00
MTB > END
K8         376.570
K6         2650.00
MTB > END
MTB >


MTB > NOTE :LET US TRY TO REDUCE THE INVENTORY
QUANTITY WHILE MEETING DEMAND.
MTB > LET K1=11400
MTB > EXEC 'MINCOST' 5 TIMES
K8         332.285
K6         750.000
MTB > END
K8         383.570
K6         3950.00
MTB > END
K8         398.427
K6         3850.00
MTB > END
K8         358.856
K6         2750.00
MTB > END
K8         430.570
K6         3950.00
MTB > END
MTB >
```

```
MTB > NOTE:LET US INCREASE OUR DELIVERY.
MTB > LET K1=11450
MTB > EXEC 'MINCOST' 5
K8        435.784
K6       4800.00
MTB > END
K8        424.070
K6       3900.00
MTB > END
K8        381.213
K6       3200.00
MTB > END
K8        381.213
K6       2500.00
MTB > END
K8        387.927
K6       2900.00


MTB > SET C3
DATA> .01 .03 .08 .20 .40 .67 .85 .93 .97 1.0
DATA> END
MTB > NAME C3 'CUMPROB'
MTB > PRINT C1 C3
 ROW   GALLONS   CUMPROB

   1      1050      0.01
   2      1150      0.03
   3      1250      0.08
   4      1350      0.20
   5      1450      0.40
   6      1550      0.67
   7      1650      0.85
   8      1750      0.93
   9      1850      0.97
  10      1950      1.00

MTB > PLOT C1 C3
        GALLONS
       2000.+                                              *
            -                                            *  *
            -                                         *  *
            -                                      *
       1600.+
            -                              *
            -                        *
            -                  *
            -           *
       1200.+        *
            -     *
            -   *
            -
            -
        800.+
            +---------+---------+---------+---------+---------+CUMPROB
          0.00      0.20      0.40      0.60      0.80      1.00
```

Refinement Of The Demand Submodel

If the demand intervals are small enough, using their midpoints as the corresponding demand level may be sufficient for many applications. However, suppose we want demand to vary "continuously" over the entire interval. In that case we may wish to use an empirical model which captures the trend of the cumulative probability distribution applying the techniques presented in Chapter 5. As presented in Exhibit 7.3.2, an analysis of a scatterplot and divided difference table suggests that a third-order polynomial should do a reasonable job of capturing the trend of the data. Using the least-squares criterion, we obtain the following cubic polynomial:

$$q_i = 1062.99 + 2271.0x_i - 4331.6x_i^2 + 2921.7x_i^3$$

where x_i is a random number between 0 and 1. Thus a daily demand can be generated by substituting a random number between 0 and 1 into the above formula. A multiplot of the above polynomial together with the observed midpoints is included in Exhibit 7.3.2.

--------------------------------- EXHIBIT 7.3.2 ---------------------------------

Simulating The Inventory Problem Using A Continuous Polynomial To Represent The Daily Demand Submodel

```
MTB > EXEC 'DDTABLE'
    COUNT   =    10.000
  ROW  GALLONS  PROBS     C200         C201         C202         C203         C204         C205

    1     1050   0.01       *            *            *            *            *            *
    2     1150   0.03     0.0002         *            *            *            *            *
    3     1250   0.08     0.0005     0.0000015        *            *            *            *
    4     1350   0.20     0.0012     0.0000035    0.0000000        *            *            *
    5     1450   0.40     0.0020     0.0000040    0.0000000   -0.0000000        *            *
    6     1550   0.67     0.0027     0.0000035   -0.0000000   -0.0000000    0.0000000        *
    7     1650   0.85     0.0018    -0.0000045   -0.0000000   -0.0000000   -0.0000000   -0.0000000
    8     1750   0.93     0.0008    -0.0000050   -0.0000000    0.0000000    0.0000000    0.0000000
    9     1850   0.97     0.0004    -0.0000020    0.0000000    0.0000000   -0.0000000   -0.0000000
   10     1950   1.00     0.0003    -0.0000005    0.0000000   -0.0000000   -0.0000000   -0.0000000

  ROW       C206         C207         C208         C209

    1         *            *            *            *
    2         *            *            *            *
    3         *            *            *            *
    4         *            *            *            *
    5         *            *            *            *
    6         *            *            *            *
    7         *            *            *            *
    8     0.0000000        *            *            *
    9    -0.0000000   -0.0000000        *            *
   10     0.0000000    0.0000000    0.0000000    0.0000000

MTB > END
MTB >
```

```
MTB > NOTE WE WILL GENERATE DATA TO PLOT A SMOOTH CURVE OF OUR CUBIC MODEL.
MTB > SET C10
DATA> .01:1.0/.01
DATA> END
MTB > LET C11=1062.99+2271*C10-4331.6*C10**2+2921.7*C10**3
MTB > NAME C11 'PREDGAL',C10 'CPROB'
MTB >
```

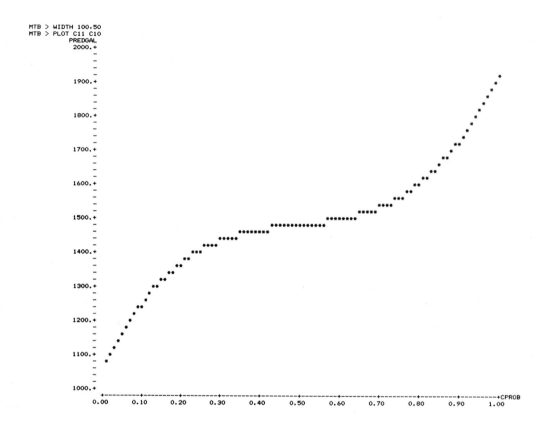

```
MTB > NOTE WE WILL NOW COMPARE THE CUBIC MODEL TO THE ORIGINAL DATA.
MTB > MPLOT C1 C3 C11 C10
    GALLONS
   2000.+
        -
        -
        -                                                              A
        -
   1900.+                                                              B
        -                                                              B
        -                                                              B
        -                                                            2
        -                                                          B
   1800.+                                                          B
        -                                                        B
        -                                                       BA
        -                                                      B
        -                                                    BB
   1700.+                                                   B
        -                                                 BB
        -                                              2
        -                                             BB
        -                                           BB
   1600.+                                          BB
        -                                        BB
        -                               A      BBB
        -                                    BBBB
        -                                  BBBBB
   1500.+                             BBBBBBBB
        -                     BBBBBBBBBBBBBBB
        -               BBBBB2BB
        -            BBBBB
        -          BBBB
   1400.+        BBB
        -       BB
        -      B2
        -     BB
        -    BB
   1300.+   BB
        -   B
        - A  B
        -   BB
   1200.+   B
        -   B
        - A B
        -   B
        -   B
   1100.+  B
        - B
        - A
        -
   1000.+
        +---------+---------+---------+---------+---------+---------+---------+---------+---------+---------+CUMPROB
       0.00      0.10      0.20      0.30      0.40      0.50      0.60      0.70      0.80      0.90      1.00

MTB > NOTE OUR MODEL APPEARS REASONABLE.
MTB >
```

7.3   EXERCISES

1.   In many situations the time T between deliveries and the order
quantity Q is not fixed.   Instead, an order is placed for a
specific amount of gasoline.   Depending on how many orders are
placed in a given time interval, the time it takes to fill an order
varies.   You have no reason to believe that the performance of the
delivery operation will change.   Therefore, you have examined your
records for the last 100 deliveries and found the "lag times" or
"extra days" (as shown below) required to fill your order.

     Construct a Monte Carlo simulation for the "lag time" submodel.
Use Minitab to test your submodel by running 100 trials and compare
the simulated number of occurrences of the various lag times with
the historical data.   Then make an additional run with 1000 trials.

191

Lag time (in days)	Number of occurrences
2	10
3	25
4	30
5	20
6	13
7	2
	100

2. In the case that a gasoline station runs out of gas, the customer will simply go to another station. However, in many other situations (name a few) some of the customers will place a "back-order." If the order is not filled within a certain time period, which varies from customer to customer in a probabilistic fashion, the customer will cancel his or her order. Suppose we examine the historical records for 1000 customers and find the data shown in the following table. That is, 200 customers will not even place an order, and 150 more customers will cancel if their order is not filled within one day, and so forth.

Hypothetical data for a backorder submodel

Number of days customer is willing to wait before cancelling	Number of occurrences	Cumulative occurrences
0	200	200
1	150	350
2	200	550
3	200	750
4	150	900
5	50	950
6	50	1000
	1000	

Construct a Monte Carlo simulation in Minitab for the backorder submodel. Test your submodel by running 1000 trials and compare the number of occurrences of the various cancellations against the historical data.

3. Incorporate the submodels developed in Problems 1 and 2 into the model we developed in the illustrative example. What is the new optimal inventory strategy? (Use the same data as given in the example.)

4. The hypothetical lifetime batting statistics for the Atlanta Braves baseball team are tabled below. Management is considering using a computer to predict the outcome of a player's turn at bat based on his past performance. Construct an algorithm that simu-lates Horner's turn at bat. Include the use of a random number generator. Construct a computer code in Minitab to simulate a

season for Horner (800 times at bat) and print out his statistics for the season. Then run your simulation for 5 seasons.

Player	At Bats	Singles	Doubles	Triples	Homeruns	Outs
Washington	2000	290	100	20	150	1440
Jackson	100	15	8	2	0	75
Perry	50	10	3	1	1	35
Murphy	3000	400	160	40	300	2100
Horner	2500	300	100	50	250	1800
Hubbard	1000	150	65	5	30	750
Ramirez	1500	250	100	15	40	1095
Benedict	2000	300	100	10	50	1540
McMurtry	100	11	2	1	1	85

8

AN INTRODUCTION TO LINEAR PROGRAMMING WITH MINITAB

```
COMMANDS USED:
        READ
        NAME
        LET
        COPY
        INVERT
        MULTIPLY
        EXECUTE
        PRINT
```

A carpenter makes tables and bookcases for a net unit profit that he estimates as $25 and $30, respectively. He is trying to determine how many of each piece of furniture he should make each week. He has up to 690 board ft of lumber to devote weekly to the project and up to 120 hours of labor. He can use the lumber and labor productively elsewhere if they are not used in the production of tables and bookcases. He estimates that it requires 20 board ft of lumber and 5 hours of labor to complete a table, and 30 board ft of lumber and 4 hours of labor for a bookcase. He also estimates that he can sell all the tables and bookcases that are produced. The carpenter wishes to determine a weekly production schedule for tables and bookcases that maximizes his profits.

Let x_1 denote the number of tables to be produced weekly, and let x_2 denote the number of bookcases. Then the model becomes

$$\text{Maximize} \quad 25x_1 + 30x_2$$

Subject to:

$$20x_1 + 30x_2 < 690 \quad \text{(Lumber)}$$

$$5x_1 + 4x_2 \leq 120 \quad \text{(Labor)}$$

$$x_1, x_2 \geq 0 \quad \text{(Nonnegativity)}$$

This model is an example of a linear program. The function $25x_1 + 30x_2$ is called the **objective function**, and x_1 and x_2 are called the **decision variables**, of the linear program.

194

An optimization problem is said to be a **linear program** if it satisfies the following properties:

---

1. There is a unique objective function.

2. Whenever a decision variable appears in either the objective function or one of the constraint functions, it must appear only as a power term with an exponent of 1, possibly multiplied by a constant.

3. No term in the objective function or in any of the constraints can contain products of the decision variables.

4. The coefficients of the decision variables in the objective function and each constraint are constants.

5. The decision variables are permitted to assume fractional as well as integer values.

A linear program has the important property that the points satisfying the constraints form a convex set. A <u>convex set</u> is one in which any two points of the set are joined by a straight line segment all of whose points lie within the set. An <u>extreme</u> point of a convex set is any point in the convex set which does not lie on a segment joining some two other points of the set. For example, the convex set for the constraints in the carpenter model is graphed and given by the polygon region ABCD in Figure 8.1. Note that there are six intersection points of the constraints, but only four of these points (namely, A, B, C, and D) satisfy <u>all</u> of the constraints and hence belong to the convex set. The points A, B, C, D are the extreme points of the polygon. The variables y_1 and y_2 will be explained further on in the sequel.

When a linear function is optimized on a convex set, the optimal solution(s), if one exists, occurs at an extreme point (or a convex combination of extreme points). Therefore, the maximum of the linear objective function $25x_1 + 30x_2$ in the carpenter problem must occur at one of the extreme points A, B, C, or D of the polygon region.

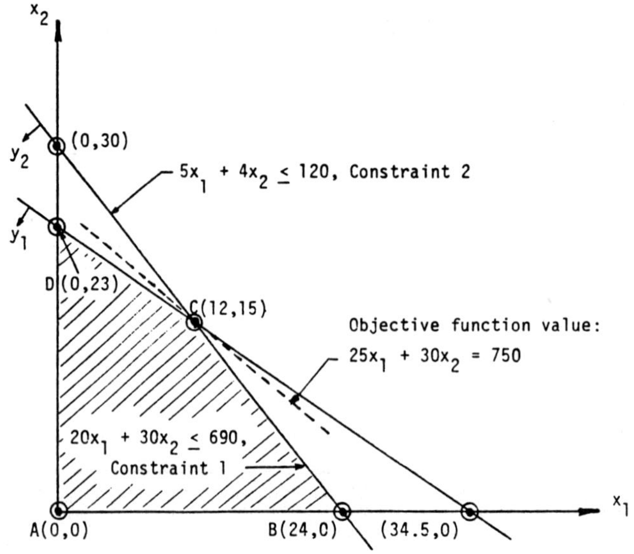
x_2
(0,30)
y_2
$5x_1 + 4x_2 \leq 120$, Constraint 2
y_1
D(0,23)
C(12,15)
Objective function value:
$25x_1 + 30x_2 = 750$
$20x_1 + 30x_2 \leq 690$,
Constraint 1
x_1
A(0,0)
B(24,0)
(34.5,0)

FIGURE 8.1 The set of points satisfying the constraints of a linear program form a convex set.

The objective function values (or profit in this case) corresponding to each extreme point are as follows:

Extreme Point	Objective Function Value
A(0,0)	\$0
B(24,0)	\$600
C(12,15)	\$750
D(0,23)	\$690

Thus the carpenter should make 12 tables and 15 bookcases each week to earn a maximum weekly profit of \$750. Use Figure 8.1 to convince yourself that the carpenter can do no better than this \$750 profit.

Finding an optimal solution

The geometrical solution to the carpenter model suggests a rudimentary procedure for finding an optimal solution to a linear program.

1. Find all the intersection points of all the constraints.

2. From the subset of intersection points that also satisfy all the constraints (i.e., that belong to the extreme points of the convex set), choose the one having the largest corresponding objective function value.

In order to implement the above procedure we must algebraically characterize the intersection points and the extreme points. Consider Figure 8.2. The convex set depicted there consists of 3 linear constraints. The nonnegative variables y_1, y_2, and y_3 indicated in the figure measure the degree by which a point satisfies each of the Constraints 1, 2, or 3, respectively. Thus, $y_2 = 0$ characterizes those points that lie precisely on Constraint 2, and a negative value for y_2 indicates the violation of Constraint 2. Likewise, the decision variables x_1 and x_2 are constrained to nonnegative values. Thus, the values of x_1 and x_2 measure the degree of satisfaction of nonnegativity of the decision variables. Note that along the x_1 axis, $x_2 = 0$. From Figure 8.2, if we set 2 of the 5 variables y_1, y_2, y_3, x_1, x_2 equal to zero simultaneously, we determine one of the intersection points. Thus all possible intersection points can be determined systematically by setting all distinguishable pairs of the 5 variables to zero and solving for the remaining 3 dependent variables. These intersection points may or may not be <u>feasible</u> solutions; that is, they may or may not belong to the convex set defined by the constraints. A negative value for one of the remaining dependent variables indicates violation of the corresponding constraint at that intersection point. For example, the intersection point where $y_2 = 0$ and $x_1 = 0$ gives a negative value for y_1 and is therefore not feasible.

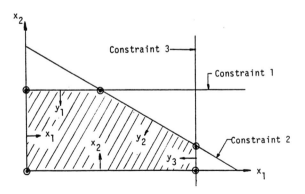

FIGURE 8.2   The variables x_1, x_2, y_1, y_2, y_3 measure the satisfaction of each of the constraints.

In general, consider a convex set consisting of m less than or equal to constraints and n decision variables x_1, x_2, ..., x_n. For each constraint i we add a slack variable y_i giving a total of m+n decision and slack variables altogether. Algebraically, this procedure yields a system of linear algebraic equations given in matrix notation by

$$AX = B$$

where A is mx(m+n), X is (m+n)x1 and B is mx1. To determine an intersection point of the constraints we choose m dependent variables among the decision and slack variables, denoted by Y, and solve

$$Y = C^{-1}B$$

where C is the mxm submatrix of A consisting of the columns of A that correspond (in the same order) to the variables in Y. We then set the n arbitrary variables to zero. Let's look at an example.

An Example

Consider again the carpenter model. Adding the nonnegative slack variables y_1 and y_2 yields the linear algebraic system

$$20x_1 + 30x_2 + y_1 \qquad = 690$$

$$5x_1 + 4x_2 \qquad + y_2 = 120$$

where

$$x_1, \ x_2, \ y_1, \ y_2 \geq 0$$

The above system can be resolved for all the intersection points using Minitab and the procedure developed for dimensional analysis in Chapter 6. First we enter the data, name the columns, and verify the input.

```
MTB > READ C1-C5
DATA> 20 30 1 0 690
DATA> 5 4 0 1 120
DATA> END
     2 ROWS READ
MTB > NAME C1 'X1',C2 'X2',C3 'Y1',C4 'Y2',C5 'RHS'
MTB > PRINT C1-C5
 ROW    X1    X2    Y1    Y2    RHS

   1    20    30    1     0     690
   2     5     4    0     1     120

MTB > COPY C1-C5 M1
```

Since there are 4 variables altogether, there are $4!/2!(2!) = 6$ intersection points (i.e., the number of combinations of 4 things taken 2 at a time). To determine an intersection point we choose 2 of the 4 variables x_1, x_2, y_1, y_2 as arbitrary, set them to 0, and then solve for the remaining 2 dependent variables by matrix inversion. Let's begin by choosing as arbitrary variables x_1 and x_2. Assigning them the value 0 gives the values for the dependent variables: $y_1 = 690$ and $y_2 = 120$. Since these values are positive, the intersection point (0,0) is feasible.

```
MTB > NOTE:WE WILL BEGIN WITH Y1 AND Y2 AS THE DEPENDENT VARIABLES.
MTB > NOTE: X1 AND X2 ARE ARBITRARY VARIABLES SET EQUAL TO ZERO.
MTB > COPY C3 C4 M2
MTB > INVERT M2 M3
MTB > MULT M3 M1 M4
MTB > COPY M4 C1-C5
MTB > PRINT C1-C5
 ROW    X1     X2     Y1    Y2     RHS

   1     20     30     1     0      690
   2      5      4     0     1      120

MTB > NOTE: INTERSECTION POINT 1 IS DEFINED WITH X1 AND X2 EQUAL TO
       ZERO.
```

For the second intersection point we choose as arbitrary variables x_1 and y_1 and set them to 0. Solving for the corresponding values of the dependent variables yields $x_2 = 23$ and $y_2 = 28$, again indicating a feasible solution. Thus the cartesian point (0,23) in Figure 8.1 is a feasible intersection point.

```
MTB > NOTE: INTERSECTION POINT 2 IS DEFINED
         WITH X2 AND Y2 AS THE DEPENDENT VARIABLES.
MTB > NOTE:X1 AND Y1 ARE EQUAL TO ZERO.
MTB > COPY M1 INTO C1-C5
MTB > COPY C2 C4 M2
MTB > INVERT M2 M3
MTB > MULT M3 M1 M4
MTB > COPY M4 C1-C5
MTB > PRINT C1-C5
 ROW        X1    X2        Y1     Y2    RHS

   1   0.66667     1    0.033333    0     23
   2   2.33333     0   -0.133333    1     28
```

For the third intersection point x_1 and y_2 are arbitrary and set to 0. Solving for the values of the dependent variables we get $x_2 = 30$ and $y_1 = -210$. Thus the first constraint is violated by 210 units at the cartesian point (0,30) so this point is infeasible.

```
MTB > NOTE: INTERSECTION POINT 3 IS DEFINED WITH
        X2 AND Y1 AS THE DEPENDENT VARIABLES.
MTB > NOTE:X1 AND Y2 ARE EQUAL TO ZERO.
MTB > COPY M1 C1-C5
MTB > COPY C2 C3 M2
MTB > INVERT M2 M3
MTB > MULT M3 M1 M4
MTB > COPY M4 C1-C5
MTB > PRINT C1-C5
  ROW        X1     X2      Y1       Y2       RHS

   1       1.25     1      0      0.25        30
   2     -17.50     0      1     -7.50      -210
```

For the fourth intersection point we choose y_1 and y_2 as arbitrary variables, set them to 0, and determine $x_1 = 12$, and $x_2 = 15$. Thus the point $(12,15)$ is feasible.

```
MTB > NOTE: INTERSECTION POINT 4 IS DEFINED WITH
        X2 AND X1 AS DEPENDENT VARIABLES.
MTB > NOTE:Y1 AND Y2 ARE EQUAL TO ZERO.
MTB > COPY M1 C1-C5
MTB > COPY C1 C2 M2
MTB > INVERT M2 M3
MTB > MULT M3 M1 M4
MTB > COPY M4 C1-C5
MTB > PRINT C1-C5
  ROW    X1    X2          Y1          Y2       RHS

   1     1     0   -0.0571429    0.428571       12
   2     0     1    0.0714286   -0.285714       15
```

The fifth intersection point is characterized by setting the arbitrary variables x_2 and y_1 to 0. The corresponding values of the dependent variables $x_1 = 34.5$ and $y_2 = -52.5$ indicate that at the point $(34.5,0)$ the second constraint is violated by 52.5 units. Hence $(34.5,0)$ is infeasible.

```
MTB > NOTE: INTERSECTION POINT 5 IS DEFINED WITH
        X1 AND Y2 AS DEPENDENT VARIABLES.
MTB > NOTE:X2 AND Y1 ARE EQUAL TO ZERO.
MTB > COPY M1 C1-C5
MTB > COPY C1 C4 M2
MTB > INVERT M2 M3
MTB > MULT M3 M1 M4
MTB > COPY M4 C1-C5
MTB > PRINT C1-C5
  ROW    X1    X2     Y1     Y2      RHS

   1     1    1.5    0.05     0      34.5
   2     0   -3.5   -0.25     1     -52.5
```

Finally, the sixth intersection poInt is determined by setting the arbitrary variables x_2 and y_2 to the value 0 and determining $x_1 = 24$ and $y_1 = 210$. Thus the point (24,0) is feasible.


```
MTB > NOTE: INTERSECTION POINT 6 IS DEFINED WITH
        X1 AND Y1 AS DEPENDENT VARIABLES.
MTB > NOTE:X2 AND Y2 ARE EQUAL TO ZERO.
MTB > COPY M1 C1-C5
MTB > COPY C1 C3 M2
MTB > INVERT M2 M3
MTB > MULT M3 M1 M4
MTB > COPY M4 C1-C5
MTB > PRINT C1-C5
```

ROW	X1	X2	Y1	Y2	RHS
1	1	0.8	0	0.2	24
2	0	14.0	1	-4.0	210

We have found the 6 intersection points by choosing in turn 2 of the 4 variables as arbitrary, assigning them the value 0, and determining values for the remaining 2 dependent variables. Of the 6 intersection points, 4 were found to be feasible and 2 infeasible:

Feasible	Infeasible
(0,0)	(0,30)
(0,23)	(34.5,0)
(12,15)	
(24,0)	

Summary of Exhaustive Procedure

Let's summarize our procedure so far. Consider a linear program with n nonnegative decision variables and m less than or equal to inequalities. For each inequality constraint adjoin a nonnegative slack variable giving a total of m+n variables. To determine an intersection point choose m dependent variables and set the remaining n arbitrary variables to 0. Thus there are $(m+n)(m+n-1)/2$ intersection points to consider (the number of combinations of m+n things taken 2 at a time). Obviously such a procedure would not be good for large problems. How then might we modify the procedure to cut down on the number of intersection points we must examine?

The Simplex Method

We are not interested in those intersection points which violate one or more of the constraints. In our notation such a violation translates to a negative value for at least one of the dependent variables at the intersection point in question. Nor are we interested in an intersection point which fails to improve the value of the objective function from its current value. The Simplex Method, developed by George Dantzig, incorporates feasibility and optimality conditions to reduce the total number of extreme points evaluated while systematically determining the optimal solution(s) to a linear program if one exists. A feasibility condition is a test which quickly determines whether a proposed intersection point is feasible. (For example, we could test to see if one of the dependent variables is negative at the intersection point.) If a proposed intersection point is feasible, it is of further interest only if it improves the current value of the objective function. An optimality condition is a test that quickly determines whether or not an intersection point corresponds to a value of the objective function better than the best value found so far.

In order to implement the Simplex Method, one must begin at some extreme point, such as (0,0) in Figure 8.1. Thus x_1, x_2 are the current arbitrary variables and y_1, y_2 are the current dependent variables. The optimality condition then determines if a variable currently assigned the value 0 (i.e., one of the current arbitrary variables) could improve the value of the objective function if it is made positive. If more than one such variable exists, one of them is chosen heuristically. For example, either x_1 or x_2 made positive in Figure 8.1 would improve the objective function value at (0,0). Thus, the optimality condition guides us in the choice of a promising variable to "enter" as a new dependent or nonzero variable. Later we give a heuristic rule for choosing the entering variable when more than one candidate exists.

The variable chosen for "entry" by the optimality condition, must replace one of the current dependent variables. The exiting variable must then be assigned the value 0 as an arbitrary variable in order to determine a new intersection point. The feasibility condition now determines which "exiting" variable the entering variable should replace. Basically, the entering variable replaces whichever exiting dependent variable can assume a zero value while maintaining nonnegative values for all the dependent variables.

The feasibility condition does not require us actually to compute the values of the dependent variables when selecting an exiting variable for replacement. Instead, you will see how we select an appropriate exiting variable by quickly determining whether any variable becomes negative if the dependent variable being considered for replacement is assigned the value 0 (a "ratio test" will be explained further on). If negativity does occur, then the dependent variable under consideration cannot be replaced. Once a set of dependent variables corresponding to a more optimal extreme point is found from the optimality and feasibility conditions, the values of the new dependent variables are determined by matrix inversion. Note that only one dependent variable is replaced at a time. Geometrically the Simplex Method proceeds from an initial extreme point to an adjacent extreme point until no adjacent extreme point is more optimal. At that time, the current extreme point is an optimal solution.

202

Many formats for implementing the above "Revised" Simplex Method exist. The format we use assumes the objective function is to be <u>maximized</u> and the constraints are <u>less than or equal to inequalities</u>. (If the problem is not expressed in this format initially, it must be redefined to be in this format.)

We first preview the Simplex Method procedure before illustrating its implementation with Minitab. Assume you wish to maximize the linear function CX subject to the constraint set AX $\leq$ B where X $\geq$ 0 and B $\geq$ 0. For illustrative purposes again consider the example:

$$\text{Maximize } 25x_1 + 30x_2$$

Subject to:

$$20x_1 + 30x_2 \leq 690$$
$$5x_1 + 4x_2 \leq 120$$
$$x_1, x_2 \geq 0$$

Format the problem

First, <u>form an augmented constraint set</u>. In our illustrative problem there are altogether four constraints, counting the two nonnegativity constraints on the decision variables. The simplex method implicitly assumes that all the variables are <u>nonnegative</u>. Thus there is no need to list the nonnegativity constraints. However, we will adjoin a constraint to ensure that any new solution we consider results in a higher value of the objective function than we have in the current solution.

Assume the current feasible solution is X_0 with objective function value CX_0. The satisfaction of the constraint $CX \geq CX_0$ ensures that a new solution X is at least as optimal as X_0. Reversing the sign of the inequality to conform to AX $\leq$ B gives $CX \leq -CX_0$. For the illustrative carpenter problem we have:

$$20x_1 + 30x_2 \leq 690$$
$$\underline{5x_1 + 4x_2 \leq 120}$$
$$-25x_1 - 30x_2 \leq -0$$

Second, <u>convert each inequality to an equality by adding a nonnegative slack variable y_i</u>. This gives the constraint set

$$20x_1 + 30x_2 + y_1 \qquad\qquad = 690$$
$$\underline{5x_1 + 4x_2 \qquad + y_2 \qquad = 120}$$
$$-25x_1 - 30x_2 \qquad\qquad + y_3 = -0$$

A feasible solution is $x_1 = x_2 = 0$, $y_1 = 690$ and $y_2 = 120$. Thus, x_1 and x_2 are arbitrary variables assigned the value 0, and y_1 and y_2 are dependent variables whose values are determined. Note that y_3 represents the current value of the objective function when the current arbitrary variables are assigned the value 0. As you will see, y_3 is really a "dummy" variable included to permit the updating of the "appended" objective function row as we compute the extreme points of the convex set by matrix inversion and multiplication. The variable y_3 is always chosen as the dependent variable occupying the last column position (third in our carpenter example) and it represents the value of the objective function.

Choose an entering variable: The optimality condition

In the above format a negative coefficient in the bottom or objective row indicates that the corresponding variable could improve the current objective function value. Thus the coefficients -25 and -30 indicate that either x_1 or x_2 could improve the current objective function value of 0. (The current constraint corresponds to $25x_1 + 30x_2 \geq 0$, with x_1 and x_2 currently 0.) When more than one candidate exists for the entering variable, a heuristic for selecting it is to choose that variable with the most negative coefficient in the objective function row. If no negative coefficients exist, the current solution is an optimal solution. In the case at hand, we choose x_2 as the entering variable. The procedure is heuristic since at this point we do not know what values the entering variables can assume.

Choose an exiting variable: The feasibility condition

The entering variable x_2 must replace either y_1 or y_2 as a dependent variable (y_3 always remains the third dependent variable). The components of the current right-hand side $[690, 120]^T$ are divided by the corresponding components of the column representing x_2, $[30, 4]^T$, to obtain the ratios $[23, 30]^T$. From the subset of ratios that are positive (both in this case) the variable corresponding to the minimum ratio is chosen for replacement (y_1 corresponding to 23 in this case). Basically, the ratios represent the value the entering variable would obtain if the corresponding exiting variable were assigned the value 0. Thus, only positive values are considered and the smallest positive value is chosen in order not to drive any variable negative. For example, if y_2 were chosen as the exiting variable assigned the value 0, x_2 would assume a value 30 as the new dependent variable, but then y_1 would be negative indicating that the intersection point (0,30) does not satisfy the first constraint. The minimum positive ratio rule illustrated above obviates enumeration of any infeasible intersection points. In the case at hand the dependent variable corresponding to the ratio 23 is y_1 so it becomes the exiting variable. Thus x_2 and y_2 become the new dependent variables.

Solve for the new dependent variables: Pivoting

Now we wish to resolve the system of equations for the values of the dependent variables y_2 and x_2 when the arbitrary variables

x_1 and y_1 are assigned the value 0. Since we have 3 rows in our matrix, we must select a 3x3 submatrix to invert. The dummy variable y_3 will always be chosen as the third dependent variable. Thus we wish to copy the columns corresponding to the variables x_2, y_2 and y_3 as a submatrix, invert the submatrix, and multiply the entire matrix by the inverted submatrix. We will use the constants K1, K2, and K3 to keep track of the current dependent variables. Other notation is defined as follows:

RHS: current right hand side value.

RATIO: ratio of current right-hand side and the column corresponding to the selected entering variable.

X1-X4: the decision variables x_1 and x_2 and the slack variables y_1 and y_2, in that order.

X5: a "dummy" column adjoined for y_3 because of the included objective function row (this variable will always be dependent).

K1: indicates the dependent variable in the first position.

K2: the dependent variable in the second position.

K3: the dependent variable in the third position.

The macro PIVOT is used to accomplish the pivot:

```
COPY CK1 CK2 CK3 M1
COPY C1-C6 M2
INVE M1 M1
MULT M1 M2 M2
COPY M2 C1-C6
PRINT C1-C6
END
```

Note that the columns corresponding to the dependent variables are copied into matrix M1 which is then inverted. Next the current matrix is copied into matrix M2. The inverted M1 then premultiplies M2 and the result is then copied back into columns C1-C6 (since the columns contain header names that are useful for interpreting the output).

After performing the pivot, the optimality condition is applied again to determine if candidate entering variables exist and to choose an appropriate one. Then the ratio test is applied to choose an exiting variable, and finally the pivot is performed. The process is repeated until no variable has a negative coefficient in the objective function row. We now use Minitab to illustrate the process for our carpenter example.

─────────────────────────── EXHIBIT 8.1 ───────────────────────────

The Revised Simplex Method Using Minitab

We begin by entering the augmented constraint matrix, naming the columns with suggestive headers, and printing the results to confirm the entry.

```
MTB > READ'LP.DATA'C1-C6
      3 ROWS READ
  ROW    C1     C2    C3    C4    C5     C6

   1     20     30     1     0     0    690
   2      5      4     0     1     0    120
   3    -25    -30     0     0     1      0

MTB > NAME C1'X1';C2'X2';C3'X3';C4'X4';C5'X5';C6'RHS';C7'RATIO'
MTB > PRINT C1-C7
  ROW    X1     X2    X3    X4    X5    RHS    RATIO

   1     20     30     1     0     0    690
   2      5      4     0     1     0    120
   3    -25    -30     0     0     1      0
```

Next we list our current dependent variables, apply the optimality condition to choose x_2 as the entering variable, and compute the ratios to determine the exiting variable by the feasibility condition:

```
MTB > LET K1=3
MTB > LET K2=4
MTB > LET K3=5
MTB > NOTE THE DEPENDENT VARIABLES ARE X3, X4, AND X5.
MTB > LET C7=C6/C2
MTB > PRINT C1-C7
  ROW    X1     X2    X3    X4    X5    RHS    RATIO

   1     20     30     1     0     0    690     23
   2      5      4     0     1     0    120     30
   3    -25    -30     0     0     1      0      0
```

Thus we choose x_2 to replace x_3 as the first dependent variable. We update K1 and perform the pivot.

```
MTB > LET K1=2
MTB > EXEC'PIVOT'
MTB > COPY CK1 CK2 CK3 M1
MTB > COPY C1-C6 M2
MTB > INVE M1 M1
MTB > MULT M1 M2 M2
MTB > COPY M2 C1-C6
MTB > PRINT C1-C6
 ROW        X1      X2        X3      X4      X5      RHS

  1    0.66667      1    0.03333       0       0       23
  2    2.33333      0   -0.13333       1       0       28
  3   -5.00000      0    1.00000       0       1      690

MTB > END
```

We apply the optimality condition to the above matrix and then choose x_1 as the entering variable. Next we apply the ratio test and choose x_4 as the exiting variable.

```
MTB > LET C7=C6/C1
MTB > PRINT C1-C7
 ROW         X1      X2         X3      X4      X5      RHS      RATIO

  1     0.66667      1     0.03333       0       0       23       34.5
  2     2.33333      0    -0.13333       1       0       28       12.0
  3    -5.00000      0     1.00000       0       1      690     -138.0
```

We note that x_1 replaces x_4 as the second dependent variable. We then update the value of $K2$ accordingly, and finally perform the pivot necessary to make x_1 a dependent variable.

```
MTB > LET K2=1
MTB > EXEC'PIVOT'
MTB > COPY CK1 CK2 CK3 M1
MTB > COPY C1-C6 M2
MTB > INVE M1 M1
MTB > MULT M1 M2 M2
MTB > COPY M2 C1-C6
MTB > PRINT C1-C6
 ROW    X1      X2        X3         X4      X5      RHS

  1     -0       1    0.071429   -0.28571      0       15
  2      1       0   -0.057143    0.42857      0       12
  3     -0       0    0.714286    2.14286      1      750

MTB > END
```

We apply the optimality condition to the above tableau and note that there are no longer any negative coefficients in the bottom row. Thus $x_1 = 12$ and $x_2 = 15$ is the optimal solution giving the objective function value $750.

In the above discussion we have not considered the case of a degenerate linear program nor any linear program where the B-vector initially contains negative values. Procedures must also be developed for handling variables which are not constrained to be nonnegative by the problem definition, and for constraints which are not less than or equal to inequalities. These and many other considerations must be provided for in a complete simplex implementation. The interactive procedure we have demonstrated here is intended solely for educational purposes.

PROJECTS

For each scenario below formulate an optimization model in the form of a linear program by answering the following questions:

A.   Identify the decision variables: What decision is to be made?

B.   Formulate the objective function: How do the decisions to be made affect the objective?

C.   Formulate the constraint set: What constraints must be satisfied? Be sure to consider whether negative values of the decision variables are allowed by the problem, and ensure that they are so constrained if required.

Use Minitab to solve each linear program by the Revised Simplex method, as presented in the text.

1.   Resource Allocation

You have just become the manager of a plant producing plastic products. Though the plant operation involves many products and supplies, you are interested in only three of the products: (1) a vinyl-asbestos floor covering, the output of which is measured in boxed lots, each covering a certain area; (2) a pure vinyl counter top, measured in linear yards; and (3) a vinyl-asbestos wall tile, measured in "squares," each covering 100 sq ft.

Of the many resources needed to produce these plastic products, you have identified four: vinyl, asbestos, labor, and time on a trimming machine. A recent inventory shows that on any given day you have 1500 pounds of vinyl and 200 pounds of asbestos available for use. Additionally, after talking to your shop foreman and to various labor leaders, you realize that you have 3 man-days of labor available for use per day and that your trimming machine is available for 1 machine-day on any given day. The following table indicates the amount of each of the four resources required to produce a unit of the three desired products, where the units are 1 box of floor cover, 1 yard of counter top, and 1 square of wall tiles. Available resources are also tabulated.

	Vinyl (lb)	Asbestos (lb)	Labor (man-days)	Machine (machine-days)	Profit
Floor Cover (per box)	30	3	.02	.01	$ 0.8
Counter Top (per yard)	20	0	.1	.05	5
Wall Tile (per square)	50	5	.2	.05	5.5
Available (per day)	1500	200	3	1	-

Formulate a mathematical model to help determine how to allocate resources in order to maximize profits.

2. Nutritional Requirements

A rancher has determined that the minimum weekly nutritional requirements for an average size horse include 40 lb of protein, 20 lb of carbohydrates, and 45 lb of roughage. These are obtained from the following sources in varying amounts at the given prices.

	Protein (lb)	Carbohydrates (lb)	Roughage (lb)	Cost
Hay (per bale)	.5	2.0	5.0	$1.80
Oats (per sack)	1.0	4.0	2.0	3.50
Feeding blocks (per block)	2.0	.5	1.0	.40
High-protein concentrate (per sack)	6.0	1.0	2.5	1.00
Requirements per horse (per week)	40.0	20.0	45.0	

Formulate a mathematical model to determine how to meet the minimum nutritional requirements at minimum cost.

3. Scheduling Production

A manufacturer of an industrial product has to meet the following shipping schedule.

Month	Required shipment (units)
January	10,000
February	40,000
March	20,000

The monthly production capacity is 30,000 units and the production cost per unit is $10. Since the company does not have a warehouse, the service of a storage company is utilized whenever needed. The storage company figures its monthly bill by multiplying the number of units in storage on the last date of the month by $3. On the first day of January the company does not have any beginning inventory. Formulate a mathematical model to assist in minimizing the sum of the production and storage costs for the three-month period.

4. Mixing Nuts

A candy store sells three different assortments of mixed nuts, each assortment containing varying amounts of almonds, pecans, cashews, and walnuts. In order to preserve the store's reputation for quality, certain maximum and minimum percentages of the various nuts are required for each type of assortment, as shown in the following table.

Assortment name	Requirements	Selling price per pound
Regular	Not more than 20% cashews Not less than 40% walnuts Not more than 25% pecans No restriction on almonds	$ .89
Deluxe	Not more than 35% cashews Not less than 25% almonds No restriction on walnuts and pecans	1.10
Blue Ribbon	Between 30% and 50% cashews Not less than 30% almonds No restriction on walnuts and pecans	1.80

The following table gives the cost per pound and the maximum quantity of each type of nut available from the store's supplier each week:

Nut type	Cost per pound	Maximum quantity available per week (lb)
Almonds	$.45	2000
Pecans	.55	4000
Cashews	.70	5000
Walnuts	.50	3000

The store would like to determine the exact amounts of almonds, pecans, cashews, and walnuts that should go into each weekly assortment in order to maximize its weekly profit. Formulate a mathematical model that will assist the store management in solving its mixing problem. Hint: How many decisions need to be made? For example, do we need to distinguish between the cashews in the Regular mix and the cashews in the Deluxe?

APPENDIX A

OBSOLETE OR NEW FEATURES IN MINITAB VERSION 85.1

```
┌─────────────────────────────────────┐
│                                      │
│   OBSOLETE COMMANDS:                 │
│             GENERATE                 │
│             CHOOSE                   │
│             OMIT                     │
│             RECODE                   │
│             SUBSTITUTE               │
│             NOPRINT                  │
│             U,N,B,I,P,D-RANDOM       │
│                                      │
│   NEW FEATURES:                      │
│             HELP Expanded            │
│             PLOT Subcommands         │
│             GPLOT                    │
│             DELETE                   │
│             CODE                     │
│             COPY Subcommands         │
│             REGRESS Subcommands      │
│             RANDOM Subcommands       │
│                                      │
└─────────────────────────────────────┘
```

Minitab has announced that Version 85.1 is now available. This version makes several commands obsolete although still functional. (Obsolete means that a new function better performs the old job.) In addition, Version 85.1 has some new functions helpful to the modeling process. These commands are listed below according to the chapter heading under which they would appear in this book. We thank Thomas Ryan for providing us with an advance copy of the Version 85.1 Reference Guide.

Chapter 1--Introduction To Minitab

The HELP command has been expanded to include additional features.

HELP (command (subcommand))

Examples. **HELP OVERVIEW, HELP COMMANDS, HELP TABLE,** OR **HELP TABLE ROWPRINTS.**

Chapter 2--Graphing Continuous Functions

a. GENERATE is now an obsolete function and is replaced by the SET command described in Chapter 2.

b. Subcommands have been introduced with the PLOT function to aid in plot control. Minitab has also developed a high resolution graphics option.

```
PLOT   Cj vs Cj

MPLOT Cj vs Cj, Cj vs Cj, ...
```

Subcommands:

```
    YINCREMENT = K
    YSTART at K (go to K)
    XINCREMENT = K
    XSTART at K (go to K)
```

High Resolution Graphics

The following commands are part of the Minitab Graphics option (consult your local computer center):

GOPTIONS

Subcommands:

```
    DEVICE = 'device'
    HEIGHT = K (inches)
    WIDTH  = K (inches)
```

```
GPLOT y in Cj vs x in Cj
```

```
GMPLOT y in Cj vs x in Cj, Cj vs Cj,...,Cj vs Cj
```

The graph plotting commands include the subcommands listed for PLOT and MPLOT as well as the following.

Subcommands:

```
    LINES (style K) connecting pairs in Cj, Cj
    CURVE (style K) connecting pairs in Cj, Cj
```

Chapter 3--Modeling Using Proportionality

a. The commands CHOOSE, OMIT, RECODE, and SUBSTITUTE are obsolete. They are replaced by **COPY** with **USE** subcommand, **COPY** with the **OMIT** subcommand, **CODE**, and **LET**, respectively.

b. The following new commands are available:

```
DELETE rows K,...,K of Cj,...,Cj
```

```
CODE (K,...,K) to K,...,K to K for data in Cj,...,Cj,
    put in Cj,...,Cj
```

```
COPY Cj,...,Cj into Cj,...,Cj
```

Subcommands:

```
USE rows K,...,K
USE rows where C = K,...,K
OMIT rows K,...,K
OMIT rows where C = K,...,K
```

Chapter 4--Model Fitting

Version 85.1 increases the number of subcommands available with REGRESS. Note that some of these subcommands used to be separate commands (e.g., NOCONSTANT).

REGRESS Subcommands:

NOCONSTANT in **EQ**uation
COEFFICIENTS put into **Cj**
MSE put into **Mj**
PREDICT function **Ej, Ej,...,Ej**

Chapter 7--Monte Carlo Simulation

a. The following random number generator commands have been replaced: URANDOM, NRANDOM, BTRIALS, BRANDOM, IRANDOM, PRANDOM, DRANDOM.

b. The NOPRINT command is no longer required because the random number generators no longer print the data.

c. The random data generators are now replaced by a single RANDOM command. Moreover, there are four new commands associated with various distributions: RANDOM generates random data from a specified distribution, PDF computes the probability density function, CDF computes the cumulative distribution function, and INVDCF computes the inverse of the distribution function. All four commands use the same subcommands to describe the distribution.

RANDOM K observations into each of **Cj,...,Cj**

This command puts a random sample of K observations into each column. The subcommand specifies which distribution you want. If no subcommand is given, random data are simulated from a normal distribution with mu = 0 and sigma = 1.

PDF for values in **Ej** (put into **Ej**)

For a discrete distribution, PDF calculates probabilities for the specified values, sometimes called the discrete probability density function. For a continuous distribution, PDF calculates the (continuous) probability density function (often called simply "the density function").

CDF for values in Ej (put into Ej)

The CDF, or cumulative distribution function, for any value x, is the probability that a random variable with the specified distribution has a value less than or equal to x. That is,

$$F(x) = P(X \leq x)$$

Sometimes the CDF is called the "distribution function."

INVCDF for values Ej (put into Ej)

The INVCDF is the inverse of the cumulative distribution function.

Subcommands:

BERNOULLI	**p** = **K**	
BINOMIAL	**n** = **K**	**p** = **K**
POISSON	**mu** = **K**	
INTEGER	**a** = **K**	**b** = **K**
DISCRETE	values in **C** probabilities in **C**	
NORMAL	**mu** = **K**	**sigma** = **K**
UNIFORM	**a** = **K**	**b** = **K**
T	**v** = **K**	
F	**u** = **K**	**v** = **K**
CAUCHY	**a** = **K**	**b** = **K**
LAPLACE	**a** = **K**	**b** = **K**
LOGISTIC	**mu** = **K**	**sigma** = **K**
LOGNORMAL	**mu** = **K**	**sigma** = **K**
CHISQUARE	**v** = **K**	
EXPONENTIAL	**b** = **K**	
GAMMA	**a** = **K**	**b** = **K**
WEIBULL	**a** = **K**	**b** = **K**
BETA	**a** = **K**	**b** = **K**

d. HISTOGRAM has been enhanced with subcommands, and the high resolution graphic basics version of GHISTOGRAM draws bars with lines instead of stars.

HISTOGRAM of column Cj,...,Cj

GHISTOGRAM of column Cj,...,Cj

Subcommands:
INCREMENT = K
START at **K** (**END** at **K**)
SAME scales for all columns

APPENDIX B

CREATING DIVIDED DIFFERENCE TABLES: USER INSTRUCTIONS, DOCUMENTATION, AND PROGRAM LISTINGS

```
DIFFERENCE TABLE:
        DDTABLE
        DRIVER
        XX
        XY
        DIV
        DIVDIV
```

USER INSTRUCTIONS

STEP 1. ENTER x in C1.

STEP 2. ENTER y in C2.

STEP 3. NAME C1 and C2, if desired.

STEP 4. Obtain a scatterplot of C2 vs C1, if desired.

STEP 5. EXECute 'DDTABLE' listed below.

There will be a pause for the computations. The length of the pause increases with the size of the data set.

STEP 6. Interpret the table qualitatively to determine if a low-order polynomial is worth investigating.

STEP 7. After choosing your low-order polynomial, if appropriate, you will want to obtain the best fit model by regression.

DOCUMENTATION

This divided difference file is set up to handle approximately 50 data points. To handle more points, change the constant assigned to K5, K7, K6, K9, and K10. Also change the dummy columns C50 and C60 to column numbers which allow the expansion of your data sets. In the program the following columns are used:

Columns

C1: x or independent variable
C2: y or dependent variable
CK3: a dummy column of x's used to calculate Δx
CK1: Δx column
CK4: a copy column of the dummy x's created for the loop sequence
CK5: a dummy column of y's used to calculate Δy
CK6: Δy
CK9: $\Delta y / \Delta x$
CK7: a copy of column CK9 used to replace CK5 or y_{n-1}
CK10: a copy of column CK9 used as y_n to calculate Δy
C50: a column of rows 1 to n+1 where n is the number of data points
C60: a column of rows 1 to n+1

Constants

The constants K1, K3, K4, K5, K7, K6, K9, K10 refer to column numbers of the CK capability which can be changed to expand the number of data entries allowed.

K2: n, the number of data points
K300: n-2, or the number of columns in the divided difference table
K400: the last column of divided differences

EXECUTION

The Driver program formulates and creates all columns to be used. It calculates K2(n), the number of data points. It creates a column of length n+1 for creating Δx and Δy, and it calculates n-2, the number of columns in the divided difference table. Dummy columns are created of length n+1 to be used in other files. This Driver program is executed only once.

The following routines are executed n-2 times in a loop:

(1) 'XX': This creates $\Delta x = x_n - x_{n-1}$. The column x_{n-1} has been previously copied in the Driver file. A missing value * is first inserted in the column and the first value of the column is deleted. This new column is then subtracted from the original x column. (Recall that any authentic command involving a * yields a *.) In order that the n-1st column not be lost in the loop it is copied into n. Thus, after incrementing, it becomes the new n-1 column.

(2) 'XY': This creates Δy. The first time through, the routine $y_n - y_{n-1}$ is calculated in the same manner as 'XX'. Subsequent iterations use the column of divided differences CK9. Dummy columns are used in the same manner for 'YY' as in 'XX' by inserting * first in the column and the 1st entry of the column and then subtracting from the current column of divided differences.

217

(3) 'DIV': This file calculates $\Delta y/\Delta x$. After $\Delta y/\Delta x$ is calculated it is copied into two other columns for use after the loop is incremented: one column to be y_n, the other to be manipulated into y_{n-1}. Thus a new y can be calculated in 'YY'.

(4) 'DIVDIV': This file executes the routines 'XX', 'YY', and 'DIV', and increments the counters K1, K3, K4, K5, K6, K7, K9, K10 for the columns.

(5) 'DD TABLE' executes 'DRIVER' once, executes the looping sequence 'DIVDIV' n-2 times, and prints the results of the divided difference tables. The NOECHO and ECHO commands enable the results to be printed without repeating all the macro commands. A pause is encountered as the program calculates the table.

THE PROGRAM

```
MTB>STORe 'DRIVER'
STOR>COPY C1 into C10
STOR>COPY C2 into C150
STOR>COPY C2 into C200
STOR>LET K1 = 3
STOR>LET K3 = 10
STOR>LET K4 = 11
STOR>LET K5 = 150
STOR>LET K7 = 151
STOR>LET K6 = 100
STOR>LET K9 = 200
STOR>LET K10 = 201
STOR>COUNt C1 put into K2
STOR>LET K2 = K2 + 1
STOR>LET K300 = K2 - 2
STOR>SET C50
STOR>1:K2
STOR>COPY C50 into C60
STOR>END


MTB>STORe 'XX'
STOR>INSERt 0 1 CK3
STOR>*
STOR>OMIT rows of K2 in C50 corresponding rows of CK3 C51 CK3
STOR>LET CK1 = C1 - CK3
STOR>COPY CK3 into CK4
STOR> END


MTB>STORe 'YY'
STOR> INSERt 0 1 CK5
STOR>*
STOR>OMIT row K2 in C60 and corr rows of CK5 C61 CK5
STOR>LET CK 6 = CK9 - CK5
STOR>END
```

```
MTB>STOR 'DIV'
STOR>LET CK9 = CK6/CK1
STOR>COPY CK9 into CK10
STOR>COPY CK9 into CK7
STOR>END


MTB>STORe 'DIVDIV'
STOR>EXEC 'XX'
STOR>EXEC 'YY'
STOR>EXEC 'DIV'
STOR>LET K1 = K1 + 1
STOR>LET K3 = K3 + 1
STOR>LET K4 = K4 + 1
STOR>LET K5 = K5 + 1
STOR>LET K6 = K6 + 1
STOR>LET K7 = K7 + 1
STOR>LET K9 = K9 + 1
STOR>LET K10 = K10 + 1
STOR>END


MTB>STORe 'DDTABLE'
STOR>NOECho
STOR>EXEC 'DRIVER'
STOR>EXEC 'DIVDIV' K300 times
STOR>LET K400 = 200 + K300
STOR>PRINt C1 C2, C200 - CK400
STOR>END
```

a. Simple one-term models (see proportionality and model fitting for commands used.)

b. Fitting an n-1 order polynomial to n data points, and cubic splines.

```
READ the following data into a K by K
   matrix Mj
INVErt matrix Mj, put inverse into Mj
MULTiply (K or Mj) by (K or Mj) put
   into Mj
```

c. stored commands and loops

The commands STORE and EXECUTE provide both a simple macro (or stored command file) capability and a simple looping capability.

```
STORe (in 'filename') the following
   Minitab commands
END of stored commands
EXECute commands (in 'filename')
   (K times)
NOEChO the commands that follow
ECHO the commands that follow
```

The CK capability: The integer part of a column may be replaced by a stored constant.

```
EXAMPLE: LET K1=5
         PRINt C1-CK1
         Prints C1-C5 since K1 = 5.
```

---

4. MODEL FITTING

a. Regressive commands

```
REGRess Cj on K predictors Cj,...,Cj
   (standardized residuals in Cj,
   fits in Cj)

Subcommands:
   NOCOnstant in equation
   WEIGhts are in Cj
   MSE put into K
   COEFficients put into Cj
   RESIduals into Cj (observed - fit)

NOCOnstant in all STEPwise and
   REGRess commands that follow
CONStant return to fitting a
   constant in STEPwise and REGRess
BRIEf output (using print code = K)
NOBrief return to default amount
   of output
```

b. Subcommands

Subcommands are used as special options or to convey additional information in some commands. To use a subcommand, place a semicolon at the end of the main command line; then type the subcommands. Start each subcommand on a new line and end each subcommand line with a semicolon, except for the last subcommand which ends with a period. If you forget the period, simply type it on the next line.

c. Column operations

```
COUNt the number of values in Cj
   (put into K)
N (number of nonmissing values) in Cj
   (put into K)
```

---

3. MODELING USING PROPORTIONALITY

a. Data entry

```
SET
READ the following data into
   Cj, Cj,..., Cj
READ the data 'file' into
   Cj,..., Cj
```

b. Data plotting

```
PLOT Cj vs Cj
```

c. Data transformations

```
ADD
SUBTract
MULTiply
DIVIde
RAISe
LOGE
LOGTEN
EXPOnential
LET E = expression
```

d. Plot the transformed data using the PLOT command.

e. Estimate the slope using the arithmetic LET command.

f. Data correcting

```
LET E = expression
SUBStitute
CODE
OMIT
```

6. DIMENSIONAL ANALYSIS

a. Matrix entry

READ what follows into K by K matrix Mj
READ data from 'filename' into K by K matrix Mj
PRINt Mj,...,Mj
COPY Cj,...,Cj into Mj
COPY Mj into Cj,...,Cj

b. Matrix operations

INVErt Mj, put into Mj
TRANspose of Mj, put into Mj
ADD Mj to Mj, put into Mj
SUBTract Mj from Mj, put into Mj
MULTiply Mj by Mj, put into Mj

c. Pick arbitrary variables

COPY columns corresponding to arbitrary variables into M1
INVErt M1, put into M2

d. COPY columns corresponding to dependent variables into M3.

e. MULTiply M3 by -1, store as M3.

f. MULTiply M2 by M3, store as M4.

7. SIMULATION

a. Random number generation

URANdom K observations, put into Cj
IRANdom K integers between K and K, put into Cj
DRANdom K observations, values in Cj, probabilities in Cj, put into Cj

b. Area under a curve

SIGNs of E, put into E

Mathematical Modeling with Minitab

Quick Reference Card

NOTATION:

Kj where j is an integer: indicates that a constant (such as 4.2) or a stored constant (such as K4) may be entered.

Cj where j is an integer: denotes a column (such as C12).

Ej where j is an integer: indicates that either a constant (such as K4 or 4.2) or a column j (such as C12) may be entered in that position.

Mj where j is an integer: denotes a matrix (such as M15).

1. GENERAL INFORMATION

HELP explains MINITAB commands
INFOrmation on stubs of worksheet
STOP ends current MINITAB session
SAVE (in 'file') a copy of the worksheet
RETRieve ('file')

2. GRAPHING CONTINUOUS FUNCTIONS

SET patterned data into Cj
SET data from a file into Cj
PRINt the values in columns Cj,...,Cj
PLOT y in Cj from K to K vs x in Cj from K to K
MPLOt Cj vs Cj, and Cj vs Cj, '...', and Cj vs Cj,
HEIGht of plots in K lines
WIDTh of plots in K spaces
WIDTh of plots K spaces, height K lines
OUTPut width is K
LET E = expression
NAME for Cj is 'name', for Cj is 'name'
SAVE (in 'file') a copy of the worksheet
RETRieve ('file')